Henry Jacob Bigelow

The mechanism of dislocation and fracture of the hip

With the reduction of the dislocations by the flexion method

Henry Jacob Bigelow

The mechanism of dislocation and fracture of the hip
With the reduction of the dislocations by the flexion method

ISBN/EAN: 9783337223823

Hergestellt in Europa, USA, Kanada, Australien, Japan

Cover: Foto ©berggeist007 / pixelio.de

Weitere Bücher finden Sie auf **www.hansebooks.com**

THE MECHANISM

OF

DISLOCATION AND FRACTURE OF THE HIP.

WITH

THE REDUCTION OF THE DISLOCATIONS

BY

THE FLEXION METHOD.

By HENRY J. BIGELOW, M. D.,

PROFESSOR OF SURGERY AND CLINICAL SURGERY IN THE MEDICAL SCHOOL OF HARVARD UNIVERSITY; SURGEON OF THE MASSACHUSETTS GENERAL HOSPITAL; MEMBER OF THE BOSTON SOCIETY FOR MEDICAL IMPROVEMENT, AND OF THE AMERICAN ACADEMY OF ARTS AND SCIENCES; HONORARY MEMBER OF THE SOCIÉTÉ ANATOMIQUE; CORRESPONDING MEMBER OF THE SOCIÉTÉ DE BIOLOGIE, ETC.

WITH ILLUSTRATIONS.

PHILADELPHIA:
HENRY C. LEA.
1869.

CONTENTS.

	PAGE
INTRODUCTION	3
ABSTRACT	4
DISLOCATION OF THE HIP	9
ANATOMY OF THE HIP.	
THE Y LIGAMENT	17
CAPSULE OF THE HIP	20
LIGAMENTUM TERES	21
OBTURATOR INTERNUS MUSCLE	21
OTHER MUSCLES	24
DISLOCATIONS	25
GENERAL REMARKS UPON REDUCTION	27
POSITION OF THE PATIENT AND SURGEON	31
THE Y LIGAMENT, WITH REFERENCE TO REDUCTION AND TO SUBSEQUENT TREATMENT	32
HOW THE LIMB IS TO BE HELD	33
CAPSULAR ORIFICE TO BE ENLARGED	33
FRACTURE OF THE NECK	35
FLEXION, EXTENSION, ADDUCTION, ABDUCTION, AND ROTATION	36
CIRCUMDUCTION	36
REGULAR DISLOCATIONS.	
DISLOCATION UPON THE DORSUM ILII	37
SIGNS	38
DORSAL DISLOCATION BETWEEN THE ROTATOR MUSCLES	43
REDUCTION OF THE DISLOCATION UPON THE DORSUM	46
DORSAL BELOW THE TENDON	58
SIGNS	62
THE MECHANISM OF ITS PRODUCTION, AND CAUSE OF ITS IRREDUCIBILITY	64
REDUCTION	67
THYROID AND DOWNWARD DISLOCATIONS	70
THYROID	70
SIGNS	70
VERTICAL DOWNWARD LUXATION	73
DISLOCATIONS NEAR THE TUBEROSITY OR PERINÆUM	74
REDUCTION	79
DISLOCATION UPON THE PUBES, AND BELOW THE ANTERIOR INFERIOR SPINE OF THE ILIUM. — DISLOCATION UPON THE PUBES.	84

CONTENTS.

DISLOCATION BELOW THE ANTERIOR INFERIOR SPINE OF THE ILIUM,
OR SUB-SPINOUS 86
    REDUCTION . . . 89
ANTERIOR OBLIQUE DISLOCATION 92
DISLOCATIONS IN WHICH THE OUTER BRANCH OF THE Y LIGAMENT
IS BROKEN. — SUPRA-SPINOUS DISLOCATION 95
    REDUCTION 99
EVERTED DORSAL DISLOCATION 100
    REDUCTION . . . 102
IRREGULAR DISLOCATIONS . . 103
    IRREGULAR UPWARD LUXATION . 105
    IRREGULAR DOWNWARD LUXATION . 105
    REDUCTION 107
SPECIAL CONDITIONS OF DISLOCATION.
    OLD DISLOCATIONS AND THEIR REDUCTION . 107
    DISLOCATION FROM HIP DISEASE 110
    DISLOCATION OF THE HIP, WITH FRACTURE OF THE SHAFT OF THE
    FEMUR 112
    SPONTANEOUS DISLOCATION 112
ANGULAR EXTENSION 115
FRACTURE OF THE NECK OF THE FEMUR.
    IMPACTED FRACTURE OF THE BASE 118
    ANATOMICAL STRUCTURE OF THE NECK OF THE FEMUR . . 120
    ROTATION 122
    SHORTENING . 123
    TRUE NECK 123
    REMARKS 125
    IMPACTED FRACTURE OF THE BASE OF THE NECK WITH INVERSION 128
    IMPACTED FRACTURE OF THE NECK OF THE FEMUR NEAR THE
    HEAD 131
    COMMINUTED FRACTURE OF THE TROCHANTERS WITHOUT IMPACTION 135
    FRACTURE OF THE NECK OF THE FEMUR RESULTING IN FALSE
    JOINT 137
    CRACK IN THE NECK OF THE FEMUR . 137
FRACTURE OF THE PELVIS 139
    FRACTURE OF THE RIM OF THE ACETABULUM 139
    FRACTURE IN WHICH THE HEAD OF THE FEMUR IS DRIVEN THROUGH
    THE ACETABULUM 142
    ASSERTED FRACTURE OF THE ACETABULUM, WITHOUT CREPITUS,
    FROM A SUPPOSED IMPOSSIBILITY OF KEEPING THE FEMUR IN
    PLACE 143
    FRACTURE OF OTHER PARTS OF THE PELVIS . . 144

INDEX 147

INTRODUCTION.

SOME of the more important points in this paper are presented in the following abstract, which may serve either as a table of contents or as a list of propositions to be established by the evidence in the text. The comparatively few published autopsies of dislocation of the hip, and the still fewer conclusive ones, may indeed be insufficient for the complete analysis of its complicated mechanism; but the deficient evidence may in a great measure be supplied by experiments upon the dead subject, where the essential conditions are identical with those of the living and etherized patient, notwithstanding what has been alleged to the contrary. The views here advanced may also be tested by the light they throw upon reported cases, of which I have carefully examined such as were accessible to me. If still deemed inconclusive, they may remain in doubt until established or confuted by further observation; but in the mean time it is certain that dislocated hips can be reduced upon the principles and by the rules laid down and explained in this paper. After reasonable attention to the subject, I confess that I can find no explanation so satisfactory as that here given.

ABSTRACT.

1st. The anterior part of the capsule of the hip joint is a triangular ligament of great strength, which, when well developed, exhibits an internal and external fasciculus, diverging like the branches of the inverted letter Y. It rises from the anterior inferior spinous process of the ilium, and is inserted into nearly the entire length of the anterior intertrochanteric line.

2d. The Y ligament, the internal obturator muscle, and the capsule subjacent to it, are alone required to explain the usual phenomena of the regular luxations.

3d. The regular dislocations are those in which one or both branches of the Y ligament are unbroken; and their signs are constant.

4th. The irregular dislocations are those in which the Y ligament is wholly ruptured; and they offer no constant signs.

5th. In the regular dislocations of the hip, the muscles are not essential to give position to the limb, nor desirable as aids in its reduction.

6th. The Y ligament will alone effect reduction and explain its phenomena, a part of those connected with the dorsal dislocations excepted.

7th. During the process of reduction, this ligament should be kept constantly in mind.

8th. The rest of the capsule, except perhaps that portion beneath the internal obturator tendon, need not be considered in reduction, if the capsular orifice is large enough to admit the head of the femur easily.

INTRODUCTION.

9th. If the capsular orifice is too small to allow easy reduction, it should be enlarged.

10th. The capsular orifice may be enlarged at will, and with impunity, by circumduction of the flexed thigh.

11th. Recent dislocations can be best reduced by manipulation.

12th. The basis of this manipulation is flexion of the thigh.

13th. This manipulation is efficient, because it relaxes the Y ligament, or because that ligament, when it remains tense, is a fixed point, around which the head of the femur revolves near the socket.

14th. The further manipulation of the flexed thigh may be either by traction or rotation.

15th. The dorsal dislocation owes its inversion to the external branch of the Y ligament.

16th. The so-called ischiatic dislocation owes nothing whatever of its character, or its difficulty of reduction by horizontal extension, to the ischiatic notch.

17th. "The ischiatic dislocation" is better named "*dorsal below the tendon*," and is easily reduced by manipulation.

18th. The flexion of the thyroid and downward dislocations is due to the Y ligament, which, in the first, also everts the limb, until the trochanter rests upon the pelvis.

19th. In the pubic dislocation, the range of the bone upon the pubes is limited by this ligament, which, in the sub-spinous dislocation also, binds the neck of the femur to the pelvis.

20th. In the dorsal dislocation with eversion, the outer branch of the Y ligament is ruptured.

21st. In the anterior oblique luxation, the head of the bone is hooked over the entire Y ligament, the limb being then necessarily oblique, everted, and a little flexed.

22d. In the supra-spinous luxation, the head of the femur is equally hooked over the Y ligament, the external branch of which is broken. The limb may then remain extended.

23d. In old luxations, the period during which reduction is possible is determined by the extent of the obliteration of the socket, the strength of the neck of the femur, and the absence of osseous excrescence.

24th. Old luxations may possibly require the use of pulleys, in order by traction to avoid any danger which might result to the atrophied or degenerated neck of the bone from rotation.

25th. Right-angled extension, the femur being flexed at a right angle with the pelvis, is more advantageous than that which has usually been employed.

26th. To make such extension most effective, a special apparatus is required.

Fractures of the Neck of the Thigh-Bone.

1st. The terms intra- and extra-capsular, applied to these fractures, have little practical significance.

2d. When a fracture near the head of the femur shows bony union, it is often impossible to say whether such a fracture was originally inside or outside of the capsular ligament.

3d. These fractures are therefore better divided, for practical purposes, into: 1st, the impacted fracture of the neck into the trochanter; 2d, other fractures of the neck.

4th. In this impacted fracture, the limb is everted, because the posterior cervical wall is almost always impacted, the anterior very rarely, and in a less degree.

5th. These conditions mainly result from the relative thickness of the two walls.

6th. While eversion is due to the rotation of the fractured bone on a hinge formed in the anterior cervical wall, shortening is generally due to the obliquity of this hinge.

7th. In a well-formed bone, the posterior and thin surface of the neck of the femur is prolonged into the cancellous structure beneath the intertrochanteric ridge, and is the true neck.

8th. The posterior intertrochanteric ridge is a buttress built upon the true neck, by which, when impacted, this ridge is sometimes split off.

DISLOCATION OF THE HIP.

THE original object of the following paper was to show, that, in dislocations of the hip, the position of the limb depends chiefly upon a ligament which has been of late years imperfectly described, and that the reduction of these dislocations should be managed accordingly. In connection with this subject, I also attempted to show how the anatomical structure of the neck of the femur leads to a common variety of fracture of that bone.

These views have been, as I believe, so well established by repeated experiments upon the dead subject, and so corroborated by current pathological phenomena, and by the mass of reported cases and autopsies, that little doubt can exist of their correctness.

Since about the year 1854–55, the four dislocations of the hip, as usually described, together with the method of reducing them by manipulation alone, have been annually shown to the classes attending the lectures at the Massachusetts Medical College. These four luxations were made in each case upon a single dead subject, which, notwithstanding the great laceration to which the capsule of the hip had been subjected, in no instance failed to exhibit, and to demonstrate in a striking manner, the appropriate and well-known attitude of each dislocation. In fact, the firm and persistent position of a joint displaced under such circumstances is quite remarkable. In these experiments, the fixed attitude of the limb was at first attributed to the muscles, which, when fully extended, are capable of considerable resistance in the dead subject, as well as in the living one; but it was supposed that the action of their complicated mechanism would hardly repay the labor of its study.

In the spring of 1861, having been led to expose a joint, the luxation of which had been the subject of a lecture, I was agreeably surprised to observe the simple action of the ligament, — a simplicity which subsequent experience has confirmed, and which strikingly explains the phenomena observed in the living subject.[1]

The dislocated joint alluded to presented on examination the following appearances.

1. *Great laceration of the muscles about the joint.*
2. *The ligamentum teres broken.*
3. *Laceration of the inner, outer, and lower parts of the capsule.*
4. *The anterior and upper parts of the capsule uninjured, and presenting a strong fibrous band, fan-shaped and forked.*

The remaining tendinous and muscular fibres about the joint being now completely divided, with the exception of the strong fibrous band above alluded to, it was found that the four commonly described dislocations of

---

[1] Of the figures accompanying this paper, those of the impacted fracture and of the Y ligament, numbered 1, 6, 7, 8, 19, 24, 25, 27, 29, 31, 46, 47, were reproduced, in the spring of 1861, from photographs made from this hip after dissection. In June, 1861, a paper upon the subject was read before the Boston Society for Medical Improvement; a second paper before the Massachusetts Medical Society, in May, 1864; another, in June, 1865, before the American Medical Association. In the present paper the rarer forms of dislocation have been added, with references to the more interesting reported cases.

the hip could still be exhibited without difficulty, and that in each of them the anterior portion of the capsular ligament, which alone remained, sufficed at once to direct the limb to its appropriate position and to fix it there.

Assuming, then, first, that each of these dislocations may be produced, and that, however much it may vary in degree, it uniformly exhibits its proper and familiar diagnostic signs, — secondly, that the anterior portion of the ligament of the capsule far exceeds in strength any other part of it, and that, on this account, it not only is less likely to be torn, but generally remains intact, — thirdly, that, when this alone remains, it is itself able to give position to the displaced limb, — and lastly, that, when it is divided, the other parts of the capsule, the muscles, and other tissues do this very imperfectly, as will be hereafter shown, — the *a priori* evidence is strong, that a luxated femur assumes its attitude chiefly in obedience to the traction of the tense fibres of this part of the ligament.

The resistance of a dislocated limb is unyielding, and unlike that of muscular action elsewhere; in illustration of which a few cases may be cited, taken almost at random from Sir Astley Cooper.[1]

"CASE XXXVIII. — .... He was bled thirty ounces in the recumbent posture, and small doses of tartarized antimony were administered, but without these means producing syncope. He was then placed upon a large table, and his pelvis fixed in the usual manner, by long jack-towels passed between the perinæum and the injured joint; the extending apparatus, composed also of a round towel, was then applied above the knee, and to it were attached weights to the amount of one hun-

[1] A Treatise on Dislocations and Fractures of the Joints. London, 1842.

dred and twelve pounds, fastened to a rope, which was rove through a pulley; to the influence of this weight he was submitted for four hours, but without any effect being produced. He was therefore then sent to Guy's Hospital. At half past seven, P.M., he was taken into the operating theatre. The pelvis was fixed by the common padded bandage, whilst to the knee was attached the circular bandage and pulleys, and gradual extension was made across the lower third of the opposite thigh for the space of twenty minutes, during which period he was given three grains of tartarized antimony in solution."

"Case L. — John Cockburn, a strong, muscular man, aged thirty-three, was admitted into Guy's Hospital on the 31st of July, 1819. While carrying a bag of sand at Hastings on the 24th of July, he slipped, and dislocated the left hip-joint. The foot on the affected side was plunged suddenly into a hollow in the road, which turned his knee inwards at the same time that his body fell with violence forwards. On the day of the accident two attempts were made to reduce the dislocation by pulleys, but without success; and on the 27th of July a third, but equally unsuccessful, trial was made, although continued for nearly an hour.

"It was found, upon examination, that the thigh was dislocated backwards into the ischiatic notch. The patient was carried into the operating theatre soon after his admission; and when two pounds of blood had been taken from him, and he had been nauseated by two grains of tartarized antimony, gradually administered, extension was made with the pulleys in a right line with the body, and the upper part of the thigh was raised, while the knee was depressed; the extension was continued at least for an hour and a half, during which time he took two grains more of tartarized antimony, by which he was thoroughly nauseated; the attempts at reduction, however, did not succeed."[1]

To a surgeon accustomed to the comparative ease with which the powerful muscles of a recently fractured thigh may be extended by a moderate effort continuously applied, these cases of enormous resistance in the reduction of a dislocated hip suggest a force more powerful and unyielding than that of muscular action. Indeed, the facility with which muscular contraction is overcome by ether, while the deformity and resistance of dislocation continue, should long ago have led to the conviction that muscular contraction is not a chief agent in this deformity.

But modern writers, with few exceptions, have adopted the theory of active or passive muscular resistance.

Sir Astley Cooper says: —

"With respect to the fixed position of the head of the femur, in the four dislocations which have been described, it is not to be considered as a mere matter of chance, but the natural result of the influence of the muscles, which draw the bone into these positions, and that therefore, under common circumstances, the condition is inevitable."[2] "The capsular ligaments, in truth, possess but little strength either to prevent dislocation or to resist the means of reduction." "The difficulty of reducing dislocations arises neither from the bones nor from the ligaments, but from the resistance which the muscles present by their contraction."[3]

[1] It is curious to remark that this case ultimately yielded, in the hands of Sir Astley, to the employment, unusual for him, of the flexion method, though combined with pulleys. In further illustration of the disadvantage of horizontal extension, let this case be compared with a similar one (dorsal below the tendon) where the reduction occupied three seconds (p. 69).

[2] Op. cit., p. 100.   [3] Op. cit., pp. 20, 21.

Dr. Nathan R. Smith recognizes muscular contraction as the chief agent in effecting both dislocation of the hip and its reduction.[1]

That similar views are still entertained by distinguished surgical authorities is shown by the following reported remarks of M. Chassaignac at a meeting of the Société de Chirurgie in 1865: "The employment of chloroform in the reduction of dislocations had convinced him (M. Chassaignac) that obstacles to reduction said to be due to other causes than muscular contraction were chimerical,"[2] — an observation that seems to have passed unchallenged.

Dr. Reid makes the following statement: "The chief impediment in the reduction of dislocation is the indirect action of muscles that are put upon the stretch by the malposition of the dislocated bone..... The limb or bone should be drawn in the direction which will relax the distended muscles."[3]

On the other hand, the theory of ligamentous resistance has been occasionally and distinctly recognized.

Boyer expresses his conviction of the importance of the ligament in this relation, but without proof.[4]

Professor Gunn maintains, in a paper[5] upon this subject, that any untorn or "undissected" portion of the capsular ligament is capable of producing the signs

---

[1] Med. and Surg. Memoirs, by Nathan Smith, M. D. Edited by Nathan R. Smith, M. D. Baltimore, 1831. pp. 166, 167.

[2] London Med. Times and Gazette, Dec. 1865. p. 661.

[3] Dislocation of the Femur on the Dorsum Ilii reducible without Pulleys or any other Mechanical Power. By William W. Reid, M. D., of Rochester. Transactions of the N. Y. Medical Society. Albany, 1852. p. 41.

[4] Traité des Maladies Chirurgicales, etc. Par M. le Baron Boyer. Paris, 1822. Tom. IV. p. 282.

[5] Luxations of the Hip and Shoulder, and the Agents which oppose their Reduction. By Moses Gunn, A. M., M. D., Prof. Surg. Univ. Michigan. Detroit, 1859.

of hip and shoulder luxation; while, since the reading of the present paper, Professor W. Busch,[1] at the Bonn Clinic, has recognized the resistance to the reduction of dislocation as ligamentous and capsular, although he fails to identify the anterior ligament as its principal seat.

There is no doubt that in luxation as well as fracture the muscles soon contract and adapt themselves to the new condition of things, so that the limb is steadied partly by the effort of the patient. In those luxations of the hip, for example, which exhibit great flexion, the muscles may thus contribute, when the patient is standing, to support the limb in a flexed position, while its own weight tends to straighten it; they may even help to convert a dislocation below the socket into one upon the dorsum, or into the foramen ovale, — or they may assist simply to reduce it. But there is no evidence that dislocations below the socket are capable of retaining their distinctive features, in an erect posture of the body, when the muscles are relaxed, as in the dead or etherized subject.

Again, some of the muscles are stretched and elongated by the luxated bone; and their passive strength under these circumstances, which is greater than might be supposed, has been well illustrated by Dr. Reid. But it is unnecessary to dwell upon the tenacity of the muscular fibre passively stretched to its extreme limit, because this extreme tension does not occur in the usual dislocations, being prevented by the ligamentous action. It may be remarked, however, that muscle is far less strong than ligament, and that the muscles about the hip, which are inserted near the head of the femur, are acted upon at great advantage by this powerful lever, and might yield, were they unsupported.

[1] Year-Book of Medicine, Surgery, etc., for 1863. Sydenham Society. London, 1864. p. 225.

Moreover, the dislocated hip can be shown equally well upon a subject in which the muscles have become soft by decomposition; and when the four classical dislocations have been produced upon a single subject, most of the muscular tissue immediately surrounding the joint will be found to have been torn away, while the rest may be divided, without materially affecting the power of the limb to illustrate these four luxations. On the other hand,— a fact which is conclusive,— if the entire capsule of the hip joint be divided, and the muscles left intact, these dislocations are but imperfectly represented.

Without denying, then, that muscular fibre exerts both an active and a passive force, it is here assumed that the muscles play but a subordinate and occasional part either in hindering reduction or in determining the character of the deformity, and that this is chiefly due to the resistance of a ligament, the power of which will presently be illustrated, and whose simple mechanism will explain the phenomena both of luxation and its reduction. Out of twenty-two recorded autopsies, while in two only is there any allusion to the rupture of the anterior portion of the capsule, in fourteen it is distinctly mentioned that it remained wholly or in part unbroken: a large proportion, considering that attention has hitherto not been directed to this point. It is not here maintained that this ligament will be found stripped clear of the remaining portion of the capsule: the comparatively few autopsies on record show that this is not the case: there is, however, reason to believe that the thinner portions owe their immunity from injury to the protection of the main ligament.

The theory here advanced recognizes the anterior portion of the capsular ligament as the exponent of the total agency of the capsule in giving position to the dislocated limb, and, what is more important, as so identified with the phenomena of luxation, that reduction must be accomplished almost wholly with reference to it. It remains for future autopsies to show, by careful examination, how far the usual phenomena either of luxation or its reduction can occur after rupture of this ligament.

THE Y LIGAMENT.

The ilio-femoral ligament, known also as the ligament of Bertin, has been usually described as reinforcing the capsule by a single fibrous band extending from the inferior iliac spine to the inner extremity of the anterior intertrochanteric line, and playing no very important part in health or injury. This ligament is more or less adherent to the acetabular prominence and to the neck of the femur; but it will be found, upon examination, to take its origin from the anterior inferior spinous process of the ilium, passing downward to the front of the femur, to be inserted fan-shaped into nearly the whole of the oblique "spiral" line which connects the two trochanters in front,—being about half an inch wide at its upper or iliac origin, and but little less than two inches and a half wide at its fan-like femoral insertion. Here it is bifurcated, having two principal fasciculi, one being inserted into the upper extremity of the anterior intertrochanteric line, and the other into the lower part of the same line, about half an inch in front of the small trochanter. The ligament thus resembles an inverted Y, which suggests a short and convenient name for it.

The divergent branches of the Y ligament are sometimes well developed, with scarcely any intervening membrane. In other cases the intermediate tissue is thicker, and requires to be slit or removed before the bands are distinctly defined; and sometimes the whole triangle is of nearly uniform thickness. In the subject first dissected, and from which the accompanying woodcut was designed, the external fasciculus of fibres was nearly as well developed as the inner band; in two other subjects it was actually wider and thicker. But as the internal and external branches exercise somewhat distinct functions,— the one being chiefly concerned in limiting the extension, the other the eversion, of the femur,— it is fair to infer that in a normal condition they would exhibit greater development than the intermediate fibres.

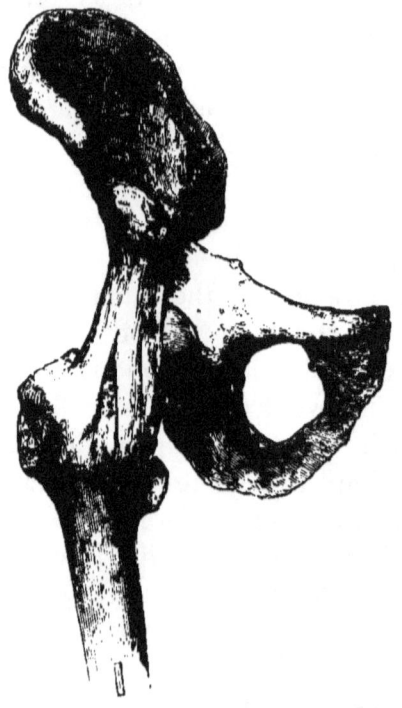

Fig. 1.

The Y ligament is of remarkable tenacity and strength, being at some points, when well developed, nearly a quarter of an inch in thickness, and forming an unyielding suspensory band, by which the femur, when in a state of extension, as in walking, is forcibly retained in its socket.

In six by no means recent subjects, taken at random from the dissecting-tables and suspended by the shoulders, the lower limbs being united to the pelvis by the Y ligament alone, this ligament

---

Fig. 1. — The Y ligament, showing its inner and outer fasciculi. The former is known as the ilio-femoral ligament, the ligament of Bertin, etc. This specimen showed the interval between the two fasciculi. (From a photograph in 1861.)

required for its rupture the attachment of weights to the foot varying in the several cases from two hundred and fifty to seven hundred and fifty pounds.[1]

The dissection of the Y ligament here represented, taken from a photograph made in 1861, first directed my attention to the anatomical arrangement and strength of its fibres. A few months ago, I found, upon referring to works in the library of my distinguished colleague, Professor O. W. Holmes, the following passages, which show that a bifurcation of this ligament was known to some of the earlier anatomists, although it has since been generally overlooked.

The first is from Winslow: —

"It [the ligament] is very thick between the anterior inferior spine of the os ilium all the way to the small anterior tuberosity which unites, as it were, the basis of the great trochanter with the basis of the neck. It is likewise very thick between the same spine and the middle part of the oblique rough line observable between the tuberosity and the little trochanter; and here likewise it is strengthened by a bundle of fibres connected to the passage of the tendon of the iliac muscle and to the inferior portion of the oblique rough line. The disposition of the ligamentous fibres of which these two thick portions are composed forms a sort of triangle with the oblique rough line which terminates the basis of the neck."[2]

[1] Although autopsies show that the whole of this ligament has sometimes been torn asunder, it may be assumed that such a lesion is likely to be of rare occurrence. Its strength probably insures its immunity in a large majority of luxations, while the constancy of their signs, which will be shown to be best explained by the action of this ligament, testifies to its integrity.

[2] An Anatomical Exposition of the Structure of the Human Body. By James Benignus Winslow. (Douglas's Translation.) London, 1776. Section II. 138, 139.

Weitbrecht, an excellent, perhaps the best, authority upon the ligaments, referring in this connection to Winslow, distinctly recognizes a forked arrangement, which he thus describes: "Partim anterius versus collum femoris et trochanterem minorem procedit, ... partim vero lateraliter versus exteriora progreditur, et circa radicem trochanteris majoris in tuberculo laterali terminatur. Atque binae hae divaricationes, una cum linea obliqua, figuram . . . . triangularem . . . . constituunt."[1]

The Webers describe the ligament as triangular, laying stress upon its thickness, which, as they assert, is greater than that of the ligament of the patella or the tendo Achillis, and concluding thus: "With this great strength, we should expect that every other part of the capsule would be ruptured before this ligament, and that even the bone itself would first yield."[2]

Capsule of the Hip.

In a front view of the cleanly dissected capsule of the joint, the Y ligament is distinctly seen, the tissue occupying its fork being sometimes a mere membrane, and sometimes thicker. The external band hinders eversion, especially when the leg is extended. Both bands limit extension. In abducting the leg, a band is raised (pubo-femoral) between the bony ridge above the thyroid foramen and the prominence at the inner part of the intertrochanteric line, hindering abduction in every position of the limb. Between this band and the Y ligament the capsule is compara-

---

[1] Syndesmologia, sive Historia Ligamentorum, etc. Josias Weitbrecht, D. M. Petropoli, 1742. p. 141.

[2] Traité d'Ostéologie, etc. S. P. Soemmerring, and G. and E. Weber. Paris, 1843. pp. 323, 324.

tively thin, and here the primitive pubic dislocation doubtless occurs. Behind and inside the pubo-femoral band, looking directly towards the thyroid foramen, is found the thinnest part of the capsule, which at this point resembles wet bladder, readily permitting the thyroid dislocation. Outside and behind the Y ligament, where the dorsal dislocations occur, the capsule is very strong, limiting adduction and rotation inward. There is also a fasciculus from the tuber ischii at its upper part to the upper part of the trochanter behind, arresting flexion and inversion. The principal ligamentous bands are the two first described, no part of the capsule comparing in strength with the Y ligament and the tissue which lies immediately behind it, beneath the tendon of the obturator internus muscle.

Ligamentum Teres.

Little can be added to the excellent paper of Mr. Struthers[1] upon the function of this ligament. When the limb is bent upon the body, it hinders eversion, thus opposing the action of the sartorius muscle, and hindering, in this position, dislocation upon the thyroid foramen. It is not, however, possessed of much strength, is ruptured in all the complete and sudden dislocations, and, according to Cruveilhier, is often undeveloped and sometimes wanting.

Obturator Internus Muscle.

It will hereafter be seen that this muscle, with which the gemelli are practically identified, is materially concerned in one variety of hip dislocation, and is important in relation to its reduction. There is an unde-

[1] Edin. Med. Journal, Nov., 1858, p. 434.

scribed peculiarity of the obturator internus which explains its strength. Its muscular belly is, in some subjects, mingled with tendinous fibres. This may be verified in dissection by reflecting the muscle from its pulley so as to expose its internal and fibrous surface. The four or five tendinous divisions which wind round the lesser sacro-sciatic notch, and which seem to disappear in the thickness of the muscular tissue, may then be traced in part to a bony origin, — some of their minute and ultimate fibres arising from the margin of the obturator foramen. The muscle, when extended, thus practically becomes a ligament, and by the attachment of its combined tissues acquires great strength. Again, the friction of the tendon over the pulley lessens the draft upon the extended muscle, and so increases its power of resistance that it is the strongest,[1] as in relation to luxation it is the most important, of the small outward rotators of the hip. That portion of the capsule which lies directly beneath the tendon is also very strong, and while their fibres mutually reinforce each other, their mechanical action in the dorsal luxations is much the same.

Arising within the pelvis, the obturator internus emerges from the pelvic cavity at a point several inches behind the great trochanter, into the back and upper part of which it is inserted. By its con-

[1] The average weight required to rupture this and the adjacent muscles in four subjects is as follows: —

Pyriformis	10 lbs.
Obturator internus	40⅔ "
Obturator externus	36⅔ "
Gluteus medius	17 "

In the only recent subject among these, the obturator internus on one side parted at 64 lbs. and on the other at 60 lbs., the obturator externus at 52 and 44 lbs., and the pyriformis at 16 lbs.

traction it draws the trochanter backward, everting the thigh, when straight, and abducting it, if flexed. Upon the dead or etherized subject, it is rendered tense in the extended limb by rotation inward, adduction being then more limited ; but in the flexed limb, and especially in extreme flexion, it is relaxed, so that, in reducing a backward dislocation, when this muscle is still entire, it might be advantageous to flex the limb as much as possible.

A curious corroboration of the importance of this muscle, as well as of the external branch of the Y ligament, is seen in a preparation[1] of my own, the case having been one of old ununited fracture of the neck of the femur in a subject, the weight of whose body in walking had been suspended chiefly between the outer branch of the Y ligament in front and the obtu-

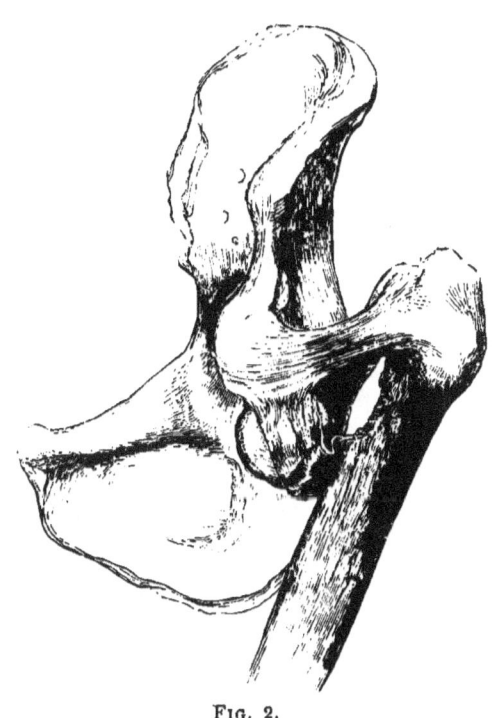

FIG. 2.

[1] No. 2715, Warren Anat. Museum, Massachusetts Medical College.

FIG. 2. — Ununited fracture of the neck of the thigh-bone, showing the hypertrophied outer fasciculus of the ligament supporting the weight of the pelvis in walking. The inner fasciculus is seen below.

DISLOCATION OF THE HIP.

Fig. 3.

rator internus behind. (Figs. 2 and 3.) This is probably the usual condition of patients after this injury, where the shaft of the femur moves freely upon the detached head of the bone.

OTHER MUSCLES.

It has already been stated that the restricted movements of the thigh in the various luxations are in part due to the active and passive resistance of several muscles which, like the psoas and iliacus, connect the femur with the pelvis and become more or less tense by its displacement; yet their action, in a practical point of view, is of secondary importance, whether considered in relation to its direction or its extent. Without the powerful ligament and the muscle already described, the regular femoral luxations would lose much of their present distinctive character; and regard being had to the action of these fibrous bands, the dislocations can be reduced with little reference to the muscles.

It may, however, be briefly stated that the gemelli are practically identified with the obturator internus, while

---

Fig. 3. — The same seen from behind, to show the tense obturator tendon bearing its share of the weight of the body; the inferior gemellus, hypertrophied, is seen below it.

the obturator externus below it and the pyriformis above it are also outward rotators, the whole forming a deep muscular layer with interstices. The quadratus femoris muscle is below the usual range of dislocations, but is easily and frequently torn; and the three glutei have comparatively little efficacy in rendering the femur immovable, even when its head is engaged, for example, beneath the medius. The psoas and iliacus exert a force in the direction of the Y ligament, especially when that is ruptured; and if the limb is elongated, the adductors, the flexors of the leg, the tensor vaginæ femoris, and the muscular fibres arising from the anterior part of the crest of the ilium, may all become more or less tense.[1]

DISLOCATIONS.

MALGAIGNE is undoubtedly right in assuming that dislocation of the hip is sometimes only partial. These various degrees of dislocation give to the limb the slight differences of position observed in different cases of the same luxation. But the observation is not new. Hippocrates, in speaking of dislocation of this joint, remarks: "In a word, luxations and sub-luxations take place in different degrees, being sometimes greater and sometimes less."[2] Yet it cannot be denied that the general character of the deformity is the same for the same dislocation, and that the phenomena were on the whole well described by Cooper, and by preceding writers from the time of Hippocrates,[3] in three or four now familiar

[1] In a case of persistent flexion after reduction, I divided these fibres. (See p. 57.)
[2] The Genuine Works of Hippocrates, etc., printed for the Sydenham Society. London, 1849. Vol. II. p. 631.
[3] Hippocrates describes the luxations on the dorsum, thyroid fora-

varieties, with three or four rarer forms of displacement, considered to be anomalous.

Accumulated experience has justified the practical value of this general division, which should not be lost sight of, either by exaggerating unimportant differences or through needlessly obscuring what is plain by names of recondite derivation. Most surgeons have seen these dislocations in the living subject, and, although the rotation, the shortening, or other displacement, may have varied a little in each case, will concede that the general position of the limb is too constant and characteristic to be slighted either as a guide to the direction of the luxation or to the force appropriate for its reduction. I have therefore adhered as far as possible to the familiar names of hip luxation, which, as usually designated, are those upon the dorsum, the ischiatic notch, the thyroid foramen, and the pubes. Great stress having been laid by most modern writers on a distinction between the two first, which, if reduced by the flexion method, are wholly unimportant variations of the same displacement, I shall endeavor to show how dorsal dislocations may be divided for practical purposes, — also, that certain other less frequent luxations, hitherto classed as anomalous, are determined by the same mechanism as the rest, and with equal certainty.

Assuming that the Y ligament exerts a uniform influence upon the several dislocations, they will be here described, with a view to their practical arrangement, according to the following classification.

men, and pubes, justly including with the first variety that which has since been called "dislocation upon the ischiatic notch," most of the cases so described by modern writers being only dorsal. In a fourth variety, the dislocation "backwards," — which has been, as I conceive, erroneously interpreted by his translators as "into the ischiatic notch," — Hippocrates describes at some length the dislocation directly downwards.

I. THE REGULAR DISLOCATIONS, in which one or both branches of the Y ligament remain unbroken.

1. DORSAL.
2. DORSAL BELOW THE TENDON *(ischiatic notch of Cooper)*.
3. THYROID AND DOWNWARD.
    Obliquely inward on the thyroid foramen, or as far as the perinæum.
    Vertically downward.
    Obliquely outward as far as the tuberosity.
4. PUBIC AND SUB-SPINOUS.
5. ANTERIOR OBLIQUE.

} *Both branches entire.*

6. SUPRA-SPINOUS.
7. EVERTED DORSAL.

} *External branch broken.*[1]

II. THE IRREGULAR DISLOCATIONS, in which the Y ligament is wholly ruptured, and whose characteristic signs are therefore uncertain.

GENERAL REMARKS UPON REDUCTION.

When the patient lies upon his back, especially if etherized, the dislocated limb gravitates, and the Y ligament becomes more and more tense as the limb approaches

[1] Although the anterior oblique, supra-spinous, and everted dorsal luxations resemble each other, it has been thought advisable to distinguish between them for the purpose of more accurately classifying recorded cases. In the anterior oblique luxation the outer branch of the Y ligament is still entire, as seen in the figure illustrating this luxation, where the ligament is of uniform thickness. This, indeed, is a form of supra-spinous luxation, but the limb cannot be brought down to a perpendicular, and corresponds in position with that in a case reported by Cooper. If the limb is forcibly brought to a perpendicular, the external branch is ruptured, and to such a case the term "supra-spinous" is here assigned. The term "everted dorsal" is intended to imply a power of eversion more or less complete. In such a case the limb may be everted at various angles, which can happen only after a rupture of the external branch of the ligament.

nearer and nearer to a state of complete extension. If, now, as is here maintained, the chief obstacle to reduction of the luxated hip is found in this ligament, it follows that the method taught by Sir Astley Cooper, the weight of whose unquestioned authority has unfortunately availed to give it currency[1] during many years, is based upon an erroneous conception of the nature of the difficulty to be encountered. By that method the limb is placed as nearly as may be in the axis of the body, thus rendering the Y ligament tense and inviting its maximum of resistance before traction is made. Hence the necessity for pulleys, the tendency of which is undoubtedly to elongate, or partly detach, at its femoral insertion, this powerful ligamentous band, at great sacrifice of mechanical force, with proportionate violence to the neighboring tissues and uncertainty as to the result. By the flexion method, which dates from a remote antiquity, the Y ligament is relaxed, its resistance annulled, and reduction often accomplished with surprising facility. The following is the statement of Hippocrates on this subject: "In some the thigh is reduced without preparation, with slight extension, directed by the hand, and with slight movement; and in some the reduction is effected *by bending the limb at the joint, with gentle shaking*."[2] In view of this observation

---

[1] See Edin. Med. Journal, May, 1867. On the Reduction of Dislocations of the Hip-Joint by Manipulation, by Thomas Annandale, Lecturer on Surgery, etc. "Its adoption in this country" (reduction by manipulation) "is as yet by no means general."

[2] Dr. Adams, in his Sydenham Translation of Hippocrates, renders this passage, "bending the limb at the joint, and *making rotation*." (Vol. II. p. 643.) Mr. Sophocles, the distinguished Professor of Greek in Harvard University, has kindly furnished me the following conclusive note, defining precisely the character of this movement.

"Your question has reference to the meaning of the word κίγκλισις, the formation of which is as follows: Κίγκλος, *wag-tail*, a well-known bird in Greece, called also σεισοπυγίς, the Latin *motacilla*. Κιγκλίζω,

of the Coan sage, 450 B. C., the indiscriminate use of pulleys hardly testifies to the progress of modern science.

Flexion lies at the foundation of success in the reduction of femoral dislocation, and compared with this the rest of the manipulation is of secondary importance. It may be taken as a safe and general rule, that, after the thigh has been flexed at a right angle, the head of the bone is to be at once guided towards the socket, and that, if the capsular orifice is large enough, the operator will in general succeed,— while it is equally certain that in the extended position of the limb the chances are all against him. When the femur is flexed, reduction may be effected in either of two ways. In the first (*traction*), the head is drawn, or forced, at once in the desired direction. In the second (*rotation*), the same result is accomplished by a rotation of the femur, which, in winding[1] the Y ligament about its neck, shortens it, and thus compels the head of the bone, as it sweeps round the socket, also to approach it. In reducing a hip, the success of rotation, adduction, abduction, and extension depends upon this ligament, while the whole manipulation must be conducted with reference to it.

In modern times the flexion method has commended itself to the good judgment of various surgeons. Many cases of successful reduction by this method are to be found in the journals, and many more

to wag (in the original sense of the term), as the bird aforesaid wags its tail.

" Κίγκλισις and κιγκλισμός, *a wagging* ; *shaking rapidly within narrow limits; gentle shaking.* The words *circumactio* and *rotation* are out of the question, — for the former is περιαγωγή, and the latter κυκλοφορία, — unless rotation be used in a peculiar sense.

" Erotian, in his Hippocratic Glossary, and Galen, define κιγκλισμός, the synonym of κίγκλισις, by βραχεῖα κίνησις, *short motion*, like that of the tail of the bird that furnishes the word."

[1] See Fig. 24.

are found, upon inquiry, to have been unrecorded. Among the papers explicitly announcing it in this country is an original one, already mentioned, published by Dr. Nathan R. Smith in 1831, and advocating manipulation against pulleys, in dislocation on the dorsum. His method, which has not been materially improved, consists of *free flexion, outward rotation, and abduction*, the employment of which he ascribes to his father, the late distinguished Professor Nathan Smith, many years before.[1] In 1852, Dr. W. W. Reid, of Rochester, N. Y., published a paper, also advocating manipulation to reduce the dorsal luxation, the only one described, by flexing the leg on the thigh, carrying the thigh across the sound limb upward over the pelvis as high as the umbilicus, and then abducting it and carrying the foot across the opposite, sound limb,[2] — a method identical in its important features with that of Nathan Smith.[3] But these and other advocates of the flexion method in this country and abroad were anticipated by Hippocrates, so far as the essential principle of flexion is concerned.

It is desirable, that, before handling the limb, the surgeon should accurately conceive not only the general object, but the details, of any movement he intends to employ, in order that the joint may not be injured by random, ill-devised, and fruitless manipulation. A slow, steady, well-directed movement will sometimes accomplish the desired result in a few seconds, while an ill-considered or uninstructed effort may be continued for a long time to no purpose.

In all the manœuvres, the gentle shaking, oscillation, or rocking motion suggested by Hippocrates may be useful as the head approaches the socket.

The following points are worthy of note.

[1] Op. cit., p. 177.   [2] Op. cit., p. 35.   [3] See also note, p. 66.

Position of the Patient and Surgeon.

The patient should be laid upon the floor, that the operator may command the limb to the best advantage, and should be etherized until the muscles of the hip are completely relaxed.[1]

---

[1] Some writers have expressed a different opinion. As a result of the theory of muscular resistance, Dr. Reid concludes that etherization to a state of complete relaxation, instead of being an advantage, is a detriment, because it prevents the contraction of the muscles required to replace the bone. (See a paper entitled "Observations on Dr. Markoe's Report of Cases of Dislocation of the Femur treated by Manipulation, by W. W. Reid, M. D., of Rochester, N. Y." New York Jour. of Med., etc., July, 1855, p. 72.) Dr. Reid seems here to add the theory of active muscular resistance to that of passive muscular resistance, already quoted from a previous paper written by him.

The British Medical Journal (Oct. 20, 1866) contains a paper read by Mr. Nunneley at a recent meeting of the British Medical Association, at Chester, "On the Reduction of Dislocation (more especially of the Hip and Shoulder) by Manipulation," in which the following views are presented: "The most important condition to be insured is a relaxed, but not perfectly helpless, flaccid, uncontractile condition of the muscles; as it is by the contraction of the muscles which are attached near to the head of the dislocated bone that reduction is mainly accomplished, ... while, on the other hand, if they be incapable of any contraction whatever, it will frequently be found to be impossible for any manipulatory movements of the surgeon to replace the bone, or, being replaced, for its being retained in that position. I feel confident that I have seen both of these causes materially interfere with success, particularly the latter one, when the muscles have been entirely paralyzed, owing to the anæsthesia having been too profound."

In 1844, a patient was made completely insensible by the administration of a bottle of port wine and half a bottle of rum in divided doses, and a hip reduction was accomplished during "a condition of muscular collapse." Lond. Med. Gaz., 1844, p. 60; from Casper's Wochenschrift, No. 9, 1844.

The Y Ligament, with Reference to Reduction and to Subsequent Treatment.

Except in the supra-spinous dislocations, the two insertions of the Y ligament are most closely approximated when the thigh is flexed upon the trunk, carried toward the navel, and rotated inward.

But it has happened, that, after unsuccessful efforts, a hip has been reduced when semi-flexed in the act of extension; which shows that in certain cases the ligament may be needlessly relaxed by extreme flexion, and may be advantageously drawn tighter by a little extension or outward rotation.[1]

I may here refer, in connection with the subsequent treatment of the patient, to the practical importance of preventing such a relaxation of the anterior ligament, whether by flexing the thigh or raising the body to a sitting posture, as may permit a recurrence of luxation. For this purpose, where the bone inclines to slip from the socket after reduction, certain cases may require not only that the limb should be kept straight, but also that the thigh should be confined for a time in the position which completed the reduction: namely, for the dorsal

---

[1] Markoe, in an interesting paper upon this subject, states that he found it necessary to vary a little from the method by flexion, abduction, and rotation outwards, recommended by Reid in dorsal dislocation. He says: "It failed us so completely from the first, that we were led to add the bringing down of the thigh to the straight position in a state of abduction, still keeping up the rocking motion; and it has been uniformly in the act of thus bringing down the limb that the reduction has been accomplished."

An Account of the Cases of Dislocation of the Femur at the Hip-Joint, treated by Manipulation alone, after the Plan proposed by W. W. Reid, M. D., of Rochester, which have occurred in the New York Hospital during the past Two Years. By Thomas M. Markoe, M. D., one of the Attending Surgeons. New York Journal of Medicine, etc., Vol. XIV., January, 1855, p. 23.

See also note, p. 48.

luxations, in abduction and eversion, or in vertical extension; for the pubic and thyroid, in a position of inward rotation and adduction: thus taking advantage of the tense ligament to bind the bone to the socket.[1]

How the Limb is to be Held.

The thigh should be bent upon the body, and the leg at a right angle with the thigh. With one hand the surgeon grasps the ankle from above, while with the other, placed beneath the head of the tibia, he lifts and guides the limb. In this way, by using the flexed leg as a lever, keeping it always flexed at a right angle for that purpose, great power is brought to bear upon the head of the femur, especially in rotating the thigh. It is therefore important to keep accurate account, during such manipulation, of the position of the head of this bone, which should not be moved at random, or indiscriminately urged when locked, lest it be broken from the shaft; and it may be convenient to remember, that, in every position, the head of the femur faces nearly in the direction of the inner aspect of its internal condyle.

Capsular Orifice to be Enlarged.

Much stress has been laid by certain writers upon the difficulty of replacing the head, when it has escaped by a small aperture in the capsule. That this condition may occasionally occur seems probable; and it is suggested by Gellé,[2] in his elaborate paper upon the subject, that, when the slit occurs close to the femoral insertion of the capsular ligament, it may be impossible to replace the head of the bone. This writer,

---
[1] See case, p. 55.
[2] Étude du Rôle de la Déchirure Capsulaire, etc. Par M. Gellé, etc. Paris, 1861.

with Malgaigne, Gunn, and others, urges the importance of placing the bone in the position it occupied when luxated, with a view to its re-entering the socket by exactly retracing its path. But while this path cannot always be known, any difficulty is easily obviated by carrying the head of the bone toward the opposite side of the socket, and thus enlarging the slit: a simple manœuvre, easily accomplished by circumducting the flexed thigh across the abdomen in a direction opposite to that in which it is desired to lead the head of the bone, which should be made in this way to pass across below the socket, and never, it is needless to say, above it, across the Y ligament. This expedient, of which I have had occasion to avail myself, will, as I believe, be in future generally adopted, when any such difficulty is encountered in reducing the hip. The subcutaneous injury is trifling, in comparison with that resulting from a protracted and ill-planned manipulation or from the brute force of pulleys. It depends for its success upon the strength of the Y ligament, which, in firmly attaching the base of the neck of the femur to the inferior spinous process of the ilium, forms a fulcrum or pivot round which the shaft and the neck of the femur can be made to revolve, like opposite spokes in a wheel; the Y ligament being strong enough to rupture, in this way, the whole of the rest of the capsule and the obturator muscle, without itself yielding.[1]

When a slit has thus been made by circumducting the neck of the bone across the posterior aspect of the capsule, the head of the bone has traversed an interval reaching in some cases from the dorsum to the thyroid foramen, and slips readily from side to side. This

[1] Reid, in the paper already quoted (New York Journal of Medicine, July, 1855, p. 69), proposes, in a similar case, "to make an incision down to the head of the bone, and open the capsular ligament sufficient to admit the return of the head into its place."

laceration already exists in most cases of dorsal dislocation below the tendon, where the head of the bone has reached a secondary position after a previous luxation downward, — and is also known to surgeons who have reduced dislocations by the old and awkward method of extension, where the bone sometimes slipped many times backward and forward from the dorsum to the foramen ovale; and yet I can find recorded only one instance of this familiar occurrence as being followed by any permanent injury, and even in that case there may have been a predisposition to the hip disease which ensued. It will hereafter be seen, that, when the head of the bone has thus been made to slip from side to side, rotation becomes a less efficient manœuvre for reduction: the bone tending at the critical moment to slip laterally away from the socket instead of into it, especially where the rim of the acetabulum is prominent or the Y ligament is relaxed. It is here that vertical traction, sudden or continued, is especially to be relied on. This will be further explained.

Fracture of the Neck.

Except in a very old subject, no apprehension need be felt of fracture from a tolerably skilful manipulation, or from circumduction with a view to tearing the capsule. The femur has, indeed, in rare instances, been fractured by manipulation as well as by pulleys;[1] and if the head of the bone be forced into a position where it is confined by the Y ligament, and

[1] Verneuil relates a case of fracture resulting from an attempt to reduce a dislocation on the pubes in a man eighty-one years of age, but only after the bone had resisted many attempts at reduction. (Med. Times and Gazette, Dec., 1865, p. 661.) Similar cases have been reported of fracture from the use both of manipulation and of pulleys. See Cooper, Op. cit., Case XXXVII.

from which it cannot escape, it will be acted upon with great power by the shaft serving as the long arm of a lever, if force be still indiscreetly applied.

Flexion, Extension, Adduction, Abduction, and Rotation.

Of these terms the two last alone require notice. If a thigh, abducted at a right angle, be rotated outward, with the knee bent, this position of the limb has been sometimes erroneously described as one of flexion. It becomes so only as the knee is brought forward.

Rotation is here always intended to apply to the thigh, the inward or outward rotation of which, in a limb bent for reduction, carries the foot in an opposite direction, and may thus lead to doubt.[1]

Circumduction.

When the patient lies on the floor, circumduction carries the knee of the dislocated limb through arcs of a horizontal circle of which the Y ligament is the centre. In attempting reduction, the direction of this motion is of primary importance, as well as the point at which it begins. The following varieties of circumduction should be distinguished from each other.

Circumduction of the extended thigh, outward, (*continued by*) " " flexed " inward.
Circumduction of the extended " inward, (*continued by*) " " flexed " outward.

The patella here always faces to the front. If it inclines outward or inward, rotation has been added to circumduction.

---

[1] "The *thigh*, abducted and a little flexed, was rotated *inward;* the *leg* and the *foot*, on the contrary, were in forced rotation *outward.*" Ollivier, Archiv. de Méd. 1823, Tom. III. p. 545.

REGULAR DISLOCATIONS.

Dislocation upon the Dorsum Ilii.

THE dorsal dislocations, having a mutual resemblance, resulting from the regular outline of the bone upon which they rest, may be more readily grouped than the others. They are found at various points of the continuously curved surface of the os innominatum, from the tuberosity of the ischium to the hollow of the ilium, — with the articular head well down beneath the external obturator muscle, or farther back in the same muscular interstice, behind the tendon of the internal obturator, — or emerging above this tendon, between it and the pyriformis muscle, — or, lastly, above the pyriformis.

The femur flexed at right angles, and thrust directly backward, tends to pass between the obturator internus and the pyriformis. At an angle of forty-five degrees it may be thrust upward and backward above the pyriformis, while in extreme flexion it is directed downward and outward toward the tuberosity and between the obturators. Inward rotation may luxate the bone at lower points, even though the limb be flexed at the same angles.

The luxation below the internal obturator tendon and the subjacent capsule is probably common, as the neck of the femur is here first arrested in its ascent from the frequent downward displacement which occurs while the limb is flexed, as it is in the great majority of such accidents. The dislocation between this muscle and the pyriformis, and the one above the pyriformis, may be assumed to be rare, the muscular fibres being probably more often torn; but future autopsies can alone show how often the other small outward rotators are lacerated[1] in the dorsal dislocation, when the internal obturator with its capsule has yielded,— or, indeed, how often the head of the bone escapes above the internal obturator tendon, and how often it ruptures it in reaching the same point.

With reference to reduction, however, these are not questions of essential importance.

The comparative infrequency of autopsies of hip dislocation, while it leaves us in the dark upon certain points which it would be desirable to have elucidated, bears evidence to the fact that this lesion is not a severe one. There can be little doubt that the small rotators near the joint are lacerated with comparative impunity, both by luxations and by their reduction. They may exert some influence upon the relative difficulty of reduction in different cases, and, as will hereafter be shown, the mutual conversion of the different varieties of dorsal luxation one into another probably depends upon the laceration of these different muscles, as well as upon that of the capsule.

Signs.

In this dislocation the limb is moderately inverted, a little shortened, and advanced. The toes cross the toes or the instep of the other foot, according to the degree of flexion and inversion, and the head of the bone may generally be felt upon the dorsum.

The inversion is chiefly due to the outer branch of

[1] In an autopsy of dorsal dislocation by Mr. Todd (see Cooper, Op. cit., Case XL.; also Dublin Hospital Reports, 1822, Vol. 3, p. 396), the dislocated extremity of the femur had occupied a large cavity between the gluteus maximus and medius; the pyriformis, gemelli, obturatores, and quadratus were completely torn across; the iliacus, psoas, and adductors were uninjured; the orbicular ligament was entire at the anterior superior parts only.

the Y ligament, as is shown by the fact that the characteristic sign disappears when this branch is divided.[1]

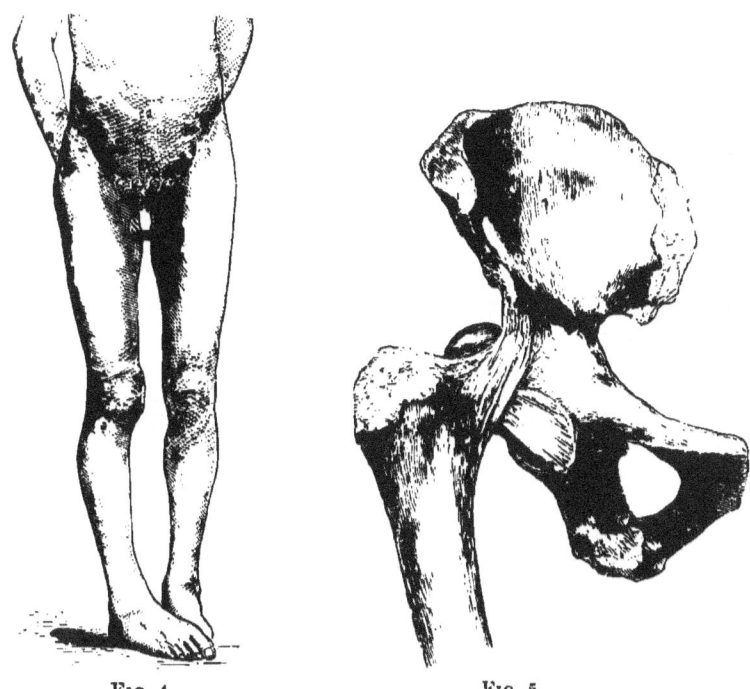

Fig. 4.   Fig. 5.

But other parts of the capsule, varying with its laceration, may assist the inversion of the limb, and when the latter is exaggerated, as when one thigh crosses the other at its middle or upper third, they may seem to be largely concerned in it: thus, if the dorsal luxation

[1] For an illustration of the condition of the anterior part of the capsule in a congenital dorsal luxation of this character, see Malgaigne (Op. cit., Pl. XXVIII. Fig. 1).

Fig. 4. — Dorsal dislocation, showing the limb inverted, the toes crossing those of the other foot. The leg has been well brought down, and exhibits but little flexion. (See p. 43.) It is often more flexed. (See Figs. 11, 13, and p. 109.)

Fig. 5. — Dorsal dislocation, showing a comparatively sound capsular ligament. For the dissected Y ligament see Figs. 6, 7.

is secondary to one below the socket, only the anterior and superior fibres will remain untorn; while, if the femur has been thrust obliquely upward and backward, attachments may be found at both the anterior and the inferior margins of the acetabulum. But it is unnecessary to consider these lesser and comparatively slender fibres. In such cases, the knee can be depressed, as indeed it often is, by the forces to which it is subjected at the time of the accident, until the exaggerated flexion and inversion have disappeared; and if even a large part of the capsule, as in the annexed figure, is stretched tense across the socket, it may then be ruptured without diminishing the inversion, which, for all practical purposes, is due to the outer branch of the Y ligament. Upon this, in fact, inversion ultimately depends; without its rupture there can be no eversion, and after its laceration the other fibres of the capsule have comparatively little strength. The rupture of the inner branch of the Y ligament does not materially change the attitude of the limb.

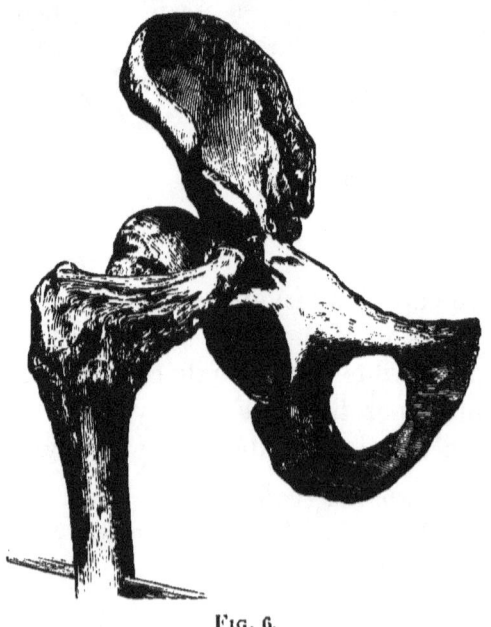

Fig. 6.

Figs. 6 and 7. — Dorsal Dislocation. These two figures are intended to show the possible range of the dissected femur when limited by the Y ligament alone. (From photographs in 1861.)

The shortening varies with the position of the head of the bone: sometimes there is little; sometimes it amounts apparently to two inches or more; but, as Malgaigne remarks, it is then complicated by flexion, and is more apparent than real.[1]

The accompanying figures (6 and 7) are intended to illustrate the operation of the Y ligament in limiting the range of the femur, and the consequent amount of shortening. At its lowest point, the head

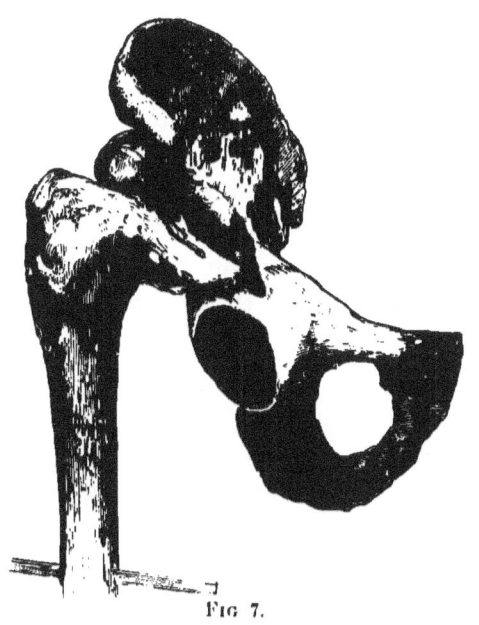

Fig 7.

of the bone corresponds to the lower part of the ischiatic notch, while it may rise upon the dorsum about an inch and a quarter above it. The former position is that most frequently occupied in dorsal luxations, the head of the bone being usually confined to the neighborhood of the socket by the unruptured muscles and by the capsular and Y ligaments (see Fig. 5). Should the femur have been thrust upwards to a higher point, it might again gravitate to the level of the ischi-

---

[1] The elongation or shortening of a dislocated thigh may, like that of the arm, be real or only apparent. If the head of the bone is luxated downwards, the limb should be longer; but flexion or abduction of the shaft approximates the usual points of measurement. To this source of error should be added the tilting of the pelvis in the femoral luxations.

atic notch, unless engaged in the interstices of the small rotators. From examination of eleven specimens of dorsal dislocation, Malgaigne[1] infers that the head of the femur generally corresponds to the ischiatic notch, and that the iliac luxation of Cooper is a pure hypothesis, while his plate illustrating it is imaginary. It will be observed that the dorsal dislocations here given in wood-cuts from photographs exhibit an inconsiderable shortening.

If, then, there be a fixed inversion of the limb with shortening, and the head of the femur is felt upon the dorsum, little desirable information is to be gained by measurement, unless in exceptional or doubtful cases, inasmuch as a primary dislocation upon the dorsum practically signifies one and the same thing, whether directly backward behind the acetabulum, or obliquely upward and backward to the full extent of the Y ligament, if this remain unbroken. A more useful indication is the degree of its mobility.

The thigh can always be flexed, and then its mobility varies with the extent of the laceration of the capsule and the adjacent tissues which bind the neck of the femur to the pelvis, — an important point, best determined by the extent to which the flexed limb can be abducted. If the bone has escaped by a large aperture in the capsule, perhaps with rupture of the obturator tendon, there will be a comparative freedom of motion and less inversion; while if the laceration is small, the movement will be restricted and the limb comparatively rigid.

By flexing and rotating the thigh, the head of the bone may be felt upon the ilium, unless the patient is very fleshy or the parts are greatly swollen; but when this sign is wanting, a differential diagnosis can be based on

---

[1] Op. cit., p. 820.

other indications. Thus, although it is practically needless to distinguish the dorsal dislocation from the one below the tendon, the latter is generally characterized by a more advanced position of the knee, the limb being more inverted and crossing the sound thigh at a higher point. On the other hand, the other regular dislocations and the fractures exhibit eversion, if we except the fracture of the neck accompanied by inversion[1] — an accident so rare that it need hardly be taken into account — and some of the fractures of the pelvis.

In the dorsal dislocation, however much the knee may be advanced and the leg inverted, even when the head of the femur is below the tendon, the thigh may be depressed by manipulation, until the knees lie almost upon the same plane; and in some cases, where, as in a female, the legs are knock-kneed, or where the knee-joint is loosely articulated, or in an old dislocation, the foot may seem not to be inverted. But the inversion of the patella, and especially the degree of resistance in everting the foot, betray the still unreduced luxation, even when the position of the limb has been much improved by efforts at reduction, which, though unsuccessful, have lacerated the capsule and loosened the muscles. This is especially true in the case of a fleshy subject, where the marks are obscure.

Dorsal Dislocation between the Rotator Muscles.

It has been said that the dorsal dislocation is often secondary, the head of the femur having first escaped below the socket. But the head of the bone may also reach the dorsum at once, by a backward thrust in

---

[1] See p 128, and also a Practical Treatise on Fractures and Dislocations, by Frank Hastings Hamilton, M. D., etc., p. 354.

the direction of its axis, which is also likely to engage the head in the muscular interstices of the rotators. Autopsies have hitherto failed to show, whether, in such a high dorsal dislocation, the internal obturator, with other inward rotators, is usually ruptured, or whether the head of the bone usually emerges above the internal obturator; but I have expressed the belief that the outward rotators are often ruptured, both by the original injury and by the protracted manipulation accompanying the use of pulleys, and that this lesion is by no means serious. It may, indeed, be difficult to distinguish between the flexion and inversion of a femur engaged above the obturator muscle and those of one in the act of ascending from the position of a luxation near the tuberosity to that below the tendon. But this distinction is practically unimportant, — since, by circumduction of the limb, a way can be cleared for the head of the bone from any point of the dorsum within the range of the Y ligament round to the thyroid foramen.

The head of the femur has been found between the obturator internus and the pyriformis, which lies above it, and has even passed still higher beneath the pyriformis, emerging between it and the gluteus minimus.[1] The bone is then drawn so far backward

[1] See an interesting case mentioned by M. Parmentier (Bulletin de la Société Anatomique, 1850, p. 177). The limb was "adducted and shortened three quarters of an inch. The femur was luxated between the pyriformis muscle and the obturator internus, the head reaching to the ischiatic spine. The button-hole thus formed opposed the reduction of the luxation in the dissected specimen." Another interesting and exceptional case has been reported by Dr. Servier. (See Bull. de la Soc. de Chir., 1863, p. 485, Report of M. Legouest.) The head of the femur, instead of escaping between the obturator internus and pyriformis muscles, as in the case described by M. Parmentier, was here found above the pyriformis, between this muscle and the gluteus minimus. The signs, so far as can be judged from the account given, were those of the

by the obturator tendon, that the outer branch of the Y becomes very tense, producing great inversion in the extended limb, and adduction when it is flexed, the obturator limiting inversion. The occur-

dorsal luxation. The capsule was ruptured posteriorly. It has been remarked, that there is great difficulty in producing these luxations in the dissected specimen, without rupturing the slender outward rotators; but if the head is made to emerge between them, either by rotation or by a direct backward thrust of the shaft, it is so embraced by the muscles, and also by the capsular orifice, which is then likely to be small, that the movements of the limb are comparatively restricted, and the muscular obstacle to their reduction may be considerable. I do not know how such cases can be identified with certainty during life. If the head of the bone has escaped by rotation of the shaft inwards, it may perhaps be reduced by outward rotation, with previous or subsequent flexion of the thigh, and thus brought to a point below the socket; although the surer way is to take the risk of rupturing all these smaller muscles by outward circumduction of the flexed limb, accompanied with outward rotation, and then to reduce the bone as usual.

M. Guersaut (Notices sur la Chirurgie des Enfants, Paris, 1864 – 67), reporting two cases which occurred under his own observation, and referring to a paper of M. Chapplain, proposes a distinction between superficial and deep iliac luxations. Such a difference would be difficult to discover in practice, either in a fleshy or a thin subject, but may have some foundation in the muscular complications just alluded to. Mr. Wormald reports a case of dislocation, *originally on the ilium*, in which, by the use of pulleys, the bone was " thrown " *upon the sciatic notch*, whence it "*could not be reduced.*" (Med. Times and Gaz., Aug. 16, 1856, p. 170.) I am at a loss to explain this case, if the facts are accurately given, except upon the hypothesis, that muscular button-holes, together with horizontal extension, determined the result.

In a case of "Dislocation of the Thigh-Bone upwards and backwards, primarily on the Dorsum Ilii, secondarily on the Sciatic Notch" (James Syme, F. R. S. E., etc., Contributions to Path. and the Pract. of Surg., p. 277), the following passage occurs : " That excellent authority, Sir Astley Cooper, though he has warned against the risk of this occurrence in reducing dislocation into the foramen ovale, has not noticed it with regard to the more common case of dislocation on the ilium." Something must be allowed for a conventional surgical prejudice against the " sciatic notch," which labors under a bad name.

rence, during life, of these varieties of dorsal luxation having been well attested, it is possible that they are comparatively common, and that many dorsal luxations occur above the internal obturator tendon,— a condition which would explain the reported "change," by pulleys, of the "dorsal" dislocation into that "upon the sciatic notch," when the limb was drawn down. The subjoined methods will accomplish their reduction before, or at any rate after, the rupture of the internal obturator; and this muscle may be ruptured at will.

The strong outer or posterior part of the capsule, which offers a resistance nearly in the direction of the obturator tendon, may also be torn by circumduction: the capsule requiring less effort for its rupture than the muscles, being inserted at the base of the neck, nearer to the centre of rotation.

REDUCTION OF THE DISLOCATION UPON THE DORSUM.

This dislocation may be equally well reduced by traction or rotation.[1]

By Traction. 1. Lay the patient, when etherized, on his back upon the floor, bend the limb at the knee, flex the thigh upon the abdomen, adduct and rotate it a little inward, to disengage the head of the bone from behind the socket. The Y ligament is then relaxed. (Fig. 8.)

If the bone can now be abducted beyond the perpendicular, the capsule and other tissues are probably so torn or relaxed that reduction may be accomplished without much difficulty; the thigh need only be forcibly

---

[1] In experiments upon the dead subject, I have twice found that the femur, after reduction by rotation, had accidentally engaged the sciatic nerve upon the front of its neck. The limb was flexed, inverted, and a little abducted, and the nerve tense and projecting in the popliteal space, as in a case of old contraction of the knee-joint. This would not have occurred if the weight of the limb had been properly sustained.

lifted or jerked towards the ceiling, with a little simultaneous circumduction and rotation outwards, to direct the head of the bone towards the socket.

2. The surgeon's foot[1] may be placed on the anterior superior spinous process of the ilium, or on the pubes, to keep the pelvis down, while he pulls the flexed knee up. Or in the same way, while assistants suspend the pelvis a few inches from the floor, by a strip of board passed transversely under the calf near the ham, the surgeon may with his foot thrust the pelvis down into its place.

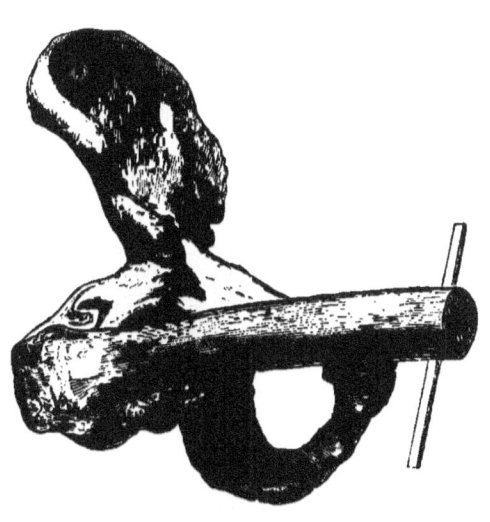

FIG. 8.

3. Flex the thigh and circumduct it outwards, across the abdomen. When it forms a large angle laterally with the trunk, the head of the bone, if it has not snapped into its place, is in or near the thyroid foramen. The rent in the capsule being thus enlarged, restore the thigh to a perpendicular, and proceed as in the last method.

4. Place the patient face downward on a table, the thigh, flexed at right angles, hanging over its edge, and bear the limb downward, with or without rotation.[2]

[1] Divested, it need hardly be said, of boot or shoe.
[2] A little girl of twelve years, upon whom six or seven attempts

FIG. 8. — Dorsal Dislocation. The result of flexion, in relaxing the Y ligament. From a photograph taken in 1861.

By Rotation. Flex the thigh and abduct or circumduct it outwards, at the same time rotating it outwards.

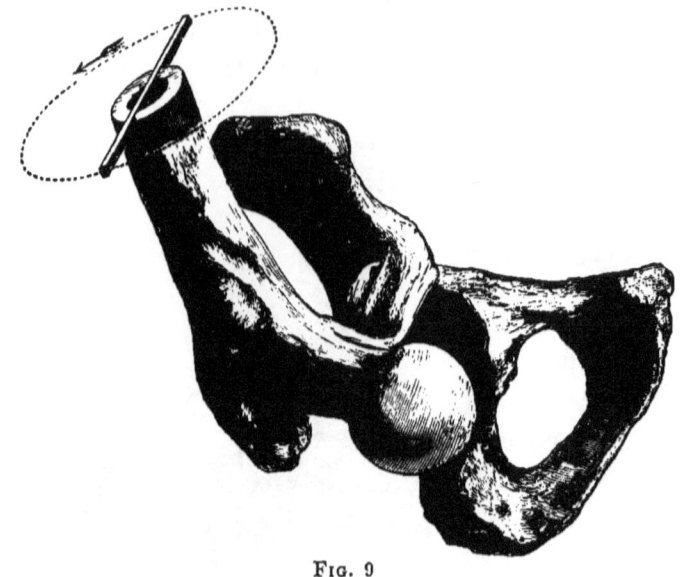

Fig. 9

The head of the bone, revolving about the great trochanter, which is fixed by the outer branch of the Y ligament, rises over the edge of the socket into its place, unless the capsule is interposed, in which case enlarge the opening, as in the third method. This is a very effective manœuvre for the reduction of the dorsal luxations, and has been described in the words, "Lift up, bend out, roll out." An imperfect comprehension of this empirical rule has led to confusion in

of an hour each had been made to reduce this luxation by straight extension, was thus placed on a board, when the head of the femur immediately slipped in. (Collin, Thèse Inaug. Montpellier, 1833.)

In 1830 Colombot had employed this method with rotation. (Documents sur la Méthode Ostéotropique. Paris, 1840.)

Fig. 9. — Dorsal Dislocation. Reduction by rotation. The limb has been flexed and abducted, and it remains only to evert it, and render the outer branch of the Y ligament tense by rotation.

its application.[1] It should be remembered, that, if the thigh is everted before it is abducted, it may be locked below the socket. For this reason it is well, especially in an old dislocation where the parts are unyielding, to invert the limb until the final abduction, when it may be everted.

When the thigh is forcibly flexed upon the abdomen, the head of the bone is lifted out from beneath the socket.[2] A little inward rotation favors the same result. If the thigh be now slowly abducted or depressed outwards, it is plain that the head of the bone, suspended by the Y ligament, must rise towards the socket, and that, when the shaft is thus abducted, outward rotation assists the entrance of the head. If the head of the bone is above the tendon of the internal obturator, this outward circumduction also ruptures

---

[1] An English journal terms this method "a knack." Mr. Cock, in a case of dorsal dislocation, gives the rule, "Lift up, bend out, roll in" (Med. Times and Gazette, June 30, 1855, p. 644), — a manœuvre which may succeed, although the method, also mentioned by Mr. Cock, of flexion, abduction, and rotation outward, is perhaps the more correct one. (Lancet, July 7, 1855, p. 6.)

The thigh being flexed, outward rotation and outward circumduction both carry the head of the bone towards the socket.

In a case of dorsal dislocation reported by Mr. W. J. Square, pulleys were applied for twenty-five minutes unsuccessfully, — and again twenty-five minutes longer, with no better result, when they were abandoned. The thigh was now flexed at right angles, and easily reduced by circumduction outwards. (Med. Times and Gazette, Nov. 13, 1858, and Am. Jour. Med. Science, Jan., 1859, p. 258.)

[2] Dr. George Sutton, of Aurora, Ind., relates a case of dorsal dislocation in which, after a failure from some cause to reduce the hip by the ordinary flexion method, a roll of cloth was placed in the groin, as a fulcrum, by which the head of the flexed femur was pried out from beneath the socket, and afterwards reduced by abduction while the limb was lifted. (Western Journal of Medicine and Surgery, Sept., 1868. Am. Journal of Med. Science, Oct., 1868, p. 588.)

the small rotator muscles. It may be needless to say, that, were the head of the bone suspended by the dissected Y ligament alone, as in some of the annexed wood-cuts, a lateral movement of the knee would perhaps cause the head of the bone to swing from side to side, instead of giving to it the desired upward tilt. This movement is hindered by the unruptured fibres of the capsule on each side of the Y ligament, which continue to a greater or less extent in the different dislocations, and contribute to the varying facility with which different cases are reduced. This is especially true of the dislocation behind the tendon of the obturator internus, where the posterior part of the capsule not unfrequently remains uninjured.

I have thus reduced the dislocated femur in living subjects by a single slow circumduction occupying from a quarter to half a minute, and also by a first rapid sweep of two seconds. The manœuvre may be perfectly accomplished without lifting the limb towards the ceiling, but is more effectual when terminated with an upward jerk.[1] If it fails, repeat the process once or twice, and then, if necessary, enlarge the opening. Or perhaps the limb is too much flexed, and the Y ligament too much relaxed. The limb may then be slowly extended from the perpendicular position, when, as the Y ligament becomes tightened, the head of the femur will rise into its place[2] (see p. 32), especially

---

[1] This upward jerk is a very efficient manœuvre, both alone and when assisted by rotation. Annibal Parea is said to have availed himself of it. "He placed the patient on his back, the pelvis being confined by assistants; he flexed the knee, raised the thigh almost vertically, grasped its lower extremity with both hands, gave it a jerk as if to raise it perpendicularly, and the luxation was instantaneously reduced." Malgaigne, Op. cit., p. 823.

[2] See a case of "ischiatic" dislocation treated in this way by Geo. W. Callender, Esq., Assistant Surgeon and Lecturer on Anatomy at

if the weight of the limb be sustained by the operator. The flexed femur is thus reduced by abduction and rotation with less flexion.[1]

If the laceration is large, and the head of the bone inclines to slip towards the thyroid foramen during abduction, this tendency is easily counteracted by the upward jerk or lift already described.

But if, upon examination, the flexed thigh cannot be abducted beyond the perpendicular, the head of the bone has either escaped by a small orifice in the capsule, which is then comparatively sound, or has also passed above the obturator or pyriformis, which are then unbroken, and is suspended just behind the edge of the socket, midway between these muscles and the Y ligament. In the former case the luxation may perhaps be reduced by flexion with abduction and outward rotation; in the latter, it is possible, but not easy, to disengage it by traction across the symphysis, the bone being lifted by a towel round the thigh at its upper part.[2] If these attempts do not succeed, the obtu-

St. Bartholomew's Hospital. (Lancet, March 14, 1868, p. 343.) Mr. Callender believes, however, that the capsule "never can offer any obstacle to the reduction of dislocations of the hip."

[1] In the extreme flexed position of the limb, the Y ligament is so relaxed that it may not afford a firm centre of rotation. (See p. 35.) A case reported by Mr. Jones (Med. Times and Gazette, April, 1856, p. 362) may serve to illustrate this point. In reducing a dorsal dislocation, the thigh was flexed as far as possible, abducted, and rotated outwards. The attempt failed; but, in gradually bringing the limb down while the same forces were applied, the head of the bone snapped into its socket when the thigh had reached a semi-flexed position. See also Mackenzie, London Hosp. Reports, 1866, Vol. III. p. 207.

[2] To dislocate the bone above the obturator tendon in the dead subject, the posterior capsule should be divided high up towards the Y ligament, and the bone then strongly flexed, adducted, and rotated outwards. By inward rotation it may be reduced.

rator muscle and the capsule can be ruptured by outward circumduction of the flexed limb, — an expedient also to be resorted to whenever the limb is especially fixed and unyielding, — after which the hip may be reduced as usual.

The following case will illustrate the method by traction. I was requested by Dr. E. A. W. Harlow, October 5, 1861, to see, with him, a stout, middle-aged Irishman, whose hip had been dislocated an hour or two before. In climbing the ladder of a freight car while the train was moving, his thigh was bent to a horizontal position just in time to be caught between this car and the next one. The flexed hip was thus dislocated backwards primarily upon the dorsum, by a force very exactly applied to the knee in front and the pelvis behind, probably with slight laceration of the capsule. The limb was shortened, the toes were firmly inverted across the instep of the other foot, the head of the femur being felt upon the dorsum. On flexion, the thigh could not be abducted as far as the perpendicular, and was unusually immovable, — the latter condition being perhaps due to the comparative integrity of the capsule. This would formerly have led to the belief that it was engaged in the ischiatic notch. It is also possible that the head of the femur may have escaped between the Y ligament and the obturator or pyriformis muscle. The patient being etherized, I flexed the limb, and made several efforts to reduce it by angular traction, but was unable to do so, — the failure being doubtless due to the small size of the capsular, and perhaps the muscular opening, which, under the same circumstances, I should now not hesitate to enlarge by circumduction. The attempt was abandoned till the next morning, when, the patient being again etherized and the limb flexed as usual, a rectangular metallic splint was applied beneath the knee, and so held by assistants as to suspend the pelvis a few inches from the floor. I then placed my foot upon the anterior superior spine of the pelvis, and at the first effort depressed the latter into its place.[1] During the patient's stay in the hospital, this limb was a little longer than the other, — an appearance I have observed in several instances, and which is perhaps due to a portion of the capsule being engaged between the head of the bone and the acetabulum, — but in 1869 it was shortened half an inch, everted, and the power of rotation impaired, apparently by dry chronic arthritis. In another case which I have lately examined, of dislocation reduced fifteen years ago by the late Dr. Hayward, this deformity of the hip from the same cause was much more strongly marked.

[1] I venture to publish the following note from Dr. Mann, of Roxbury, in illustration of the above manœuvre.

"ROXBURY, Jan. 16, 1867.
"PROFESSOR H. J. BIGELOW, —

"DEAR SIR: I take pleasure in sending you the following brief account of a case of dislocation of the right femur upon the dorsum ilii, in which I used the method for its reduction pointed out by you at the annual meeting of the Massachusetts Medical Society, in May, 1862. At that time you demonstrated a ligament described as the Y ligament, and the part performed by it in giving position to the limb and in preventing the return of the bone to its socket, together with the best means of overcoming that resistance. I was much surprised at the ease with which the reduction was accomplished, ... for I am sure in no other way could it have been accomplished with so much ease to the patient and to myself.

"I was called, July 10, 1862, to James Stump, a stout, muscular man, about fifty years of age, who, while picking cherries, lost his hold and fell from the tree, a distance of about twenty feet, to the ground. He complained of great pain in his hip, and was incapable of rising. He was picked up and conveyed to his home (a distance of three miles), where I saw him about an hour after the accident.

"I found him lying on a mattress on the floor. The right leg

The subject of the following case of dislocation of four weeks' standing was sent to me by Dr. Thomas, of Scituate.

D. P. II., æt. fifty. Four weeks ago a large door, weighing half a ton, fell upon the patient, dislocating his left hip. An irregular practitioner etherized him, and with the assistance of two men drew the leg down and told him that it was reduced. The left leg is now two inches shorter than the right, the foot inverted over the right instep, the trochanter higher and more prominent than it should be, and the head of the bone felt upon the dorsum ilii.

The reduction was effected as follows. The patient

was two inches shorter than the left, with the toes resting upon the opposite instep, the knee and foot turned inwards, and a little advanced upon the other. The limb could be bent upon the other, but could not be moved outwards. The trochanter major could be felt near the anterior superior spinous process of the ilium, and the head of the bone moving upon the dorsum ilii during rotation of the knee inwards. He was just in the position I desired, and I determined to try your method of reduction.

"Having etherized him, I placed my left foot (the boot having been removed for that purpose) upon the pelvis of the right side, and bending the leg of the patient upon the thigh, and the thigh upon the pelvis, thus relaxing the Y ligament, and placing my left arm under the knee, and grasping the ankle with the right hand, I had perfect command of the limb.

"Keeping the pelvis firmly fixed with the foot, I made a firm and pretty forcible extension with the left arm, and with a slight rotatory movement with the right hand the bone instantly slipped into its socket with a smart snapping noise which could be distinctly heard by every one in the room.

"The patient was able to walk about his room and resumed his work (which was that of fancy-basket maker) in two days. I met him upon the street three weeks after the accident, and he assured me he could walk as well as ever, saying that he had walked five miles that afternoon without fatigue.

"Very truly yours,

"BENJAMIN MANN."

was etherized and laid upon the floor. The thigh was slowly flexed upon the abdomen, and then moved laterally, to loosen the tissues about the joint. It was then returned towards the perpendicular, and jerked upwards, with a little simultaneous abduction and rotation outwards, but without success. Recognizing the comparatively untorn or reuniting capsule as the cause of the failure of this effort, I slowly circumducted the flexed thigh outwards, until the head of the bone was carried from the dorsum nearly to the thyroid foramen. After the capsular orifice was thus enlarged and the head of the bone replaced below the socket, the first upward jerk reduced the dislocation, — the whole manipulation having occupied scarcely a minute and a half.

The following cases were reduced by rotation.

In the first case the reduction was easy, and occurred in the wards of Dr. Cabot, who kindly submitted the case to me.

A man (aged twenty-four) had his left hip dislocated by the caving in of a bank of earth. The usual signs of dislocation on the dorsum were presented. To reduce it, the thigh was flexed to a perpendicular; and, in order to enlarge the capsular orifice, it was then slowly abducted with a little rotation outwards, during which it snapped into its place. The manœuvre occupied scarcely ten seconds.

It will be observed that this movement is equally suitable for extending the capsular laceration in the direction of the thyroid foramen, or, if the laceration be already sufficient, for prying the head of the bone into the socket, with the aid of the Y ligament as a fulcrum.

The following was a case of dorsal dislocation of eight months' standing, which had occurred in consequence of a fall on the floor. The patient, a woman twenty-seven years of age, had remained in bed for several months, and afterwards walked with great difficulty. The limb then presented the usual signs of dorsal dislocation, and was reduced by flexion, abduction, and eversion. I first saw her sixteen days after this operation, when the bone had again become displaced. The limb was an inch or more shorter than its fellow, and though its patella looked directly forward, and the foot was not inverted, yet the latter could not be everted like that of the sound limb, and the head of the bone was felt near the sciatic notch. By forcible flexion, abduction, and eversion, I brought the head of the bone into the socket with a snap, but, when the limb was again extended, a very slight inversion sufficed to reproduce the dislocation: in fact, the limb could not be trusted to itself. After the bone had thus repeatedly slipped out, the patient was placed in bed on her back, and the dislocation again reduced by flexion, abduction, and eversion, which brought the flexed thigh and knee down to the mattress on their outer side. The knee was then tied to the bedstead in this position by a towel, and the foot secured to the knee of the sound side, until the socket should be excavated by absorption. In two weeks she was allowed to sit up, and in two weeks more was discharged, well.[1]

In the following case of dorsal dislocation of the hip of six weeks' standing, after reduction a muscle was subcutaneously divided.

The patient, while driving a railroad hand-car, was thrown upon the track in front of it, the car passing over his body. On examination under ether, the head of the femur was felt "near the sciatic notch." After the thigh was flexed and rotated to break up the old adhesions, the dislocation was reduced by flexion, ab-

[1] Mass. Gen. Hospital, Surg. Records, Vol. 133, May, 1868.

duction, and extension. Eight days after this operation the bone had again slipped out, and at that time I first saw the patient, and made the following record in the Hospital books (Vol. 132, Aug., 1867) : —

"In the recumbent position the limb is flexed at an angle of about 40°, shortened the length of the patella, but not inverted. The trochanter is very prominent, the head of the femur being movable upon the dorsum. The dislocation is dorsal, but without inversion. The knee cannot be depressed without raising the loins. The patient, when erect, can bear about ten pounds' weight on the limb, which can be brought down by the side of the other, if the pelvis be laterally tilted to make up for the shortening, and thrown out behind to compensate the flexion. The buttock is flattened and widened as in hip disease. The feet can be everted equally, each to an angle of about 45°."

At the close of the above examination, the bone was brought into the socket by flexion, abduction, and vertical extension, though it easily slipped out of place. The next day, as the record states,. "the limb is found to be less flexed, and the head of the bone is in the socket. There is still, however, a widening and flattening of the nates on the affected side, showing that the thigh is displaced laterally, as if the socket were partially occluded, although engaging the head of the femur, while the knee is still raised about four inches above its fellow, the tensor vaginæ femoris being quite tense. The knee can be depressed, but is flexed by some elastic force, rising again." Under these circumstances, it was decided to divide the last-named muscle subcutaneously near the anterior superior spinous process, which, when done, allowed a considerable, though not complete, extension of the thigh. The limb was now brought nearly straight, and placed in a Desault's splint until the socket should be excavated by absorption. This extension was continued until September 8th, when the patient began to sit up; on the 13th he was moving about on crutches, and on the 23d he left the Hospital, there being no lengthening of the leg, and only some atrophy of the muscles of the thigh. That the luxation was unequivocal in this case is attested by the presence of the head of the bone upon the dorsum, — the femur being flexed, although the foot was straight. If the bones were sound, this absence of inversion would indicate rupture of the outer fasciculus of the Y ligament. But the marked lateral displacement, resulting from the inability of the bone fairly to enter its socket, even when placed and held there, implies some anomaly, — either the callus of fracture, the remains of capsule, or the presence of cicatricial tissue, partially occluding the socket.

A little girl three and a half years old entered the Massachusetts General Hospital with unequivocal signs of dorsal dislocation of twelve days' standing. I flexed and abducted the limb, rotating it outwards, and, after some little effort, by pressing the head of the bone towards the socket, between the fingers applied to the superior spinous process and the thumb upon the trochanter, succeeded in reducing it.

Dorsal below the Tendon.

It has been before remarked, that, when the flexion method is universally adopted, it will be practically needless to classify separately the dorsal luxations. Their varying relation to the small rotator muscles has, however, been already shown, and the strength of one of these muscles may entitle it to separate consideration.

The dislocation hitherto distinguished as " upon the

ischiatic notch," and unnecessarily associated with it, is characterized by Sir Astley Cooper as differing from the dorsal displacement chiefly in producing less shortening of the limb. I believe that no dislocation upon the ischiatic notch is worthy of the name, that no satisfactory or practical result can be based upon this distinction alone, and that it is also an error to suppose that during reduction the femur ever notably "slips into the sciatic notch,"[1] or that the sciatic notch ever offers any obstacle to its reduction. A little more or less shortening and a varying degree of inversion depend both on the position occupied by the head of the femur upon the dorsum and on the degree of laceration of the capsule. In cases of this variety which have been recorded, the signs were intrinsically the same, and reduction, if by pulleys, was usually effected in one and the same way, — unless we seek an exception to this statement in a slight variation of the angle of traction, quite as likely to occur in one case as another, and even to vary much in different attempts upon the same patient.

But there is one remarkable feature in some of the recorded instances of "ischiatic" dislocation. They were erroneously supposed to have been "irreducible." Sir Astley Cooper says, "The reduction of this dislocation is in general extremely difficult."[2] And this has thrown a shadow of uncertainty over a large number of other cases, where the observer, being persuaded that the reduction was more difficult than usual, or the limb less shortened, has taken it for granted that the head

---

[1] See, for a recent statement of this erroneous notion, Holmes's Surgery, Vol. II. p. 644 : — "That, in our attempts to reduce a dislocation upwards [*on the dorsum*], the head of the bone may slip into the sciatic notch, there is abundant evidence."

[2] Op. cit., p. 73.

of the bone was engaged in the sciatic notch and forcibly detained there, but which were in reality simple dorsal dislocations.

In view of these facts, I propose to separate the dorsal dislocations into two varieties, of some practical importance in relation to their reduction. The first has already been considered. The second includes only those cases in which the head of the femur is engaged behind the internal obturator tendon and the capsule lying beneath it, and which sometimes absolutely require the flexion method for their reduction. This is also a secondary dislocation, in which the bone, by a movement of more or less inversion, reaches its final position behind the tendon after occupying one below it, and is doubtless of frequent occurrence, as this is the point at which the luxations below the socket are first arrested in their ascent upon the dorsum. I have ventured to call it, for simplicity, *Dorsal below the tendon*, because, although the head of the femur lies behind the tendon, it is below it, also, and not above it. (See p. 43.)

The following are classical examples of this accident.

The first is from Sir Astley Cooper.

"CASE XLIII. A boy sixteen years old had a dislocation of the thigh into the foramen ovale; he was placed upon his sound side, and an extension of the superior part of his thigh was made perpendicularly; the surgeon then pressed down the knee, but the thigh being at that moment advanced, the head of the bone was thrown backwards, and passed into the ischiatic notch, from which situation it could not be reduced." It was probably this case that led Sir Astley to enjoin "great care," in reducing the thyroid luxation, "not to advance the leg in any considerable degree, otherwise the head of the thigh-bone will be forced behind the acetabulum into the ischiatic notch, from whence it cannot afterwards be reduced: this accident," he says, "I once saw happen."[1] In other words, by flexing the thigh, the Y ligament was relaxed, and the head of the bone was allowed to descend below the socket,[2] where there was an equal chance, whether, in again extending the limb, the head would return inward to the thyroid foramen, or slip outward upon the dorsum behind the obturator tendon, as actually happened.

A second is from Malgaigne.[3]

A laborer, thirty-eight years of age, had dislocated his hip backward. The next day, Lisfranc, with eight assistants, endeavored to reduce it by straight extension. At the end of an hour their efforts were abandoned, the patient being collapsed. He died on the eleventh day, of suppurative inflammation of the hip, resulting, doubtless, from the operation. At the autopsy, the bone was found to lie behind the obturator tendon, and was easily reduced by flexion.[4]

---

[1] Op. cit., p. 63.

[2] This movement is identical with that elsewhere described in connection with the three downward luxations. A similar result of relaxing the Y ligament by flexing the thigh occurred in a case of Verneuil, whose patient dislocated his hip a second time, fifteen days after the original accident, by suddenly rising to a sitting posture. The same thing happened also to a patient of Malgaigne (Malgaigne, Op. cit., p. 840), and is not uncommon.

[3] Op. cit., Pl. XXVI.

[4] For two autopsies of this dislocation see the cases of M. Bidard (Malgaigne, Op. cit., p. 835). In both these cases, of which the second seems to have been a more complete luxation than the first, the obturator internus was intact. In Queen's case, the sciatic nerve was engaged upon the neck of the femur. (Med. Chir. Trans., 1868, Vol. XXXI. p. 338.) For a case in which the head of the bone had escaped just below the socket, and was arrested there on its way toward the obturator tendon, see Ollivier, Archiv. Gén. de Méd., 1823, Tom. III. p. 545; also Lenoir, quoted by Malgaigne, Op. cit., p. 873.

In an interesting case, reported with its autopsy by Thomas Wormald (London Med. Gaz., 1837, Vol. XIX. p. 657), the dislocated

Signs.

The distinctive signs of this dislocation, dorsal below the tendon, may be thus stated.

The limb is extremely inverted. It crosses the opposite thigh, even as high as the middle of it, although in

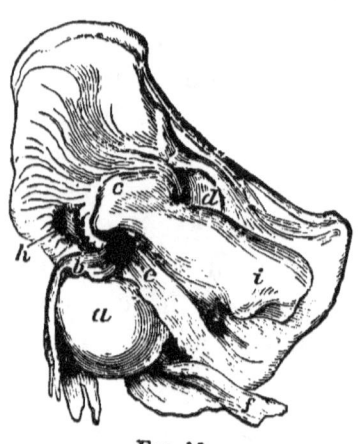

Fig. 10.

limb was shortened and inverted, forming about half a right angle with the body, while the shaft of the femur, crossing the symphysis pubis, was fixed immovably in this situation. The head of the femur had escaped above the quadratus, through a rent of the capsule opposite the upper part of the tuber ischii, compressing the sciatic nerve, and had plunged beneath the obturator externus muscle so as to engage this muscle upon its anterior face. The obturator internus was completely ruptured; the pyriformis and gemelli were partially so; also the gluteus medius and minimus at their posterior edge. In this case, the head of the bone, escaping between the two muscles, had passed forward beneath the external obturator, instead of retreating backward behind the tendon of the internal obturator. The luxation probably occurred when the limb was flexed and extremely inverted,

---

Fig. 10. — Wormald's case. Copied from London Med. Gaz., 1837. The head of the bone, *a*, is seen engaged beneath *e*, the obturator externus muscle. *f*, sciatic nerve; *b*, obturator internus; *c*, *i*, trochanters; *d*, socket; *h*, gluteus.

Fig. 11. — Dislocation below the tendon. The inversion is here seen to be greater than in the common dorsal luxation, and would be still further exaggerated in the recumbent position.

Fig. 12. — Profile view of the same, showing the leg advanced.

Fig. 13. — Dislocation downward and outward towards the tuberosity. This may be considered a first step to luxation behind the tendon, which it inclines to become when the patient is upright. The limb may occupy any interval between these two luxations, the quadratus muscle readily yielding. See Fig. 10.

DISLOCATION BELOW THE TENDON.

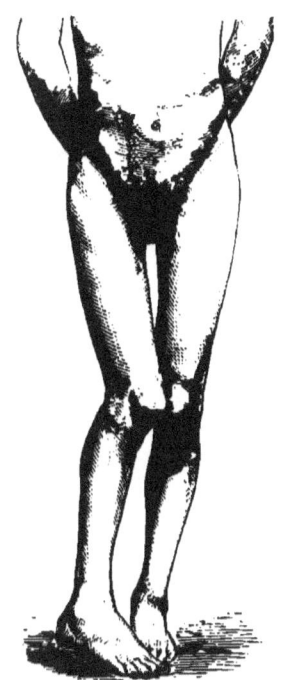

Fig. 11.

Fig. 12.

Fig. 13.

the upright position it may gravitate to a lower point. It is considerably in advance of the sound limb. By manipulation, the capsular and muscular fibres may be so relaxed or torn that this dislocation may be made to resemble one higher up on the dorsum, or be actually converted into one by rupture of the obturator muscle.

The Mechanism of its Production, and Cause of its Irreducibility.

In this luxation, the bone first escapes below the socket, or on its thyroid aspect, when the thigh is flexed, as it generally is. The limb being extended by subsequent violence, while the neck of the bone is unyieldingly suspended beneath the socket by the Y ligament, the head slips upward, not only behind the acetabulum, but also behind the capsule and the internal obturator muscle. The fibres of the latter, instead of lying transversely behind the head, as when in place (Fig. 14), now lie obliquely

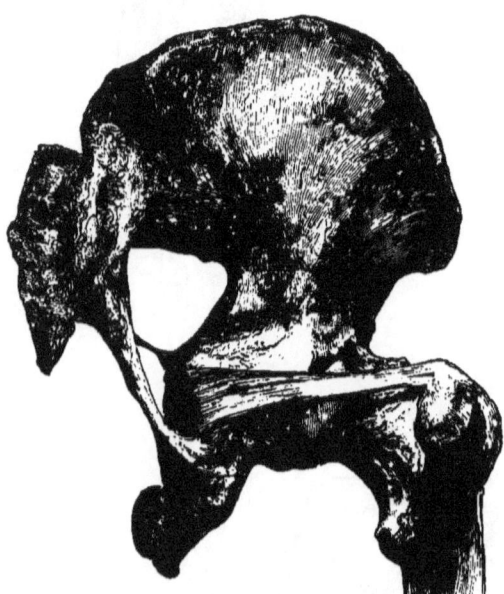

Fig. 14.

to which position may also be referred the rupture of the obturator internus muscle. By depressing the knee, the head of the bone would have been carried upward and backward, and the laceration so extended

in front of it, a tendinous wall, interposed between it and the acetabulum (Fig. 15.)[1]

The difficulty of reducing this luxation by extension in the axis of the body will be readily understood. In the absence of both posterior capsule and internal obturator muscle, the head of the bone might be slipped forward over the lower margin of the

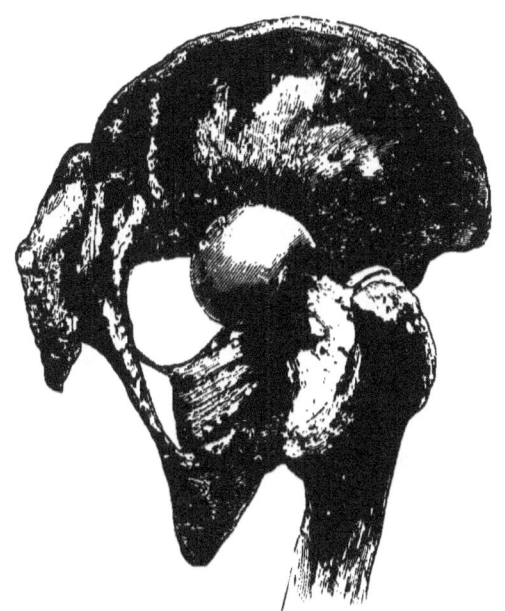

FIG. 15.

that reduction would have been easy. If the flexed knee had been circumducted outward, the external obturator muscle would have been partially ruptured; and this lesion probably occurs, when the head of the bone is carried from the dorsum to the thyroid foramen, or *vice versa*. The regular thyroid luxation, however, occurs above this muscle, the upper edge of which only need then be ruptured. The external obturator and much of the quadratus are torn in the common dorsal dislocation.

[1] A case of dislocation behind the tendon, with fracture of the

FIGS. 14 and 15. — The mechanism of the dorsal dislocation below the tendon.

Fig. 14 shows the head of the bone in its socket, with the obturator tendon in its natural position behind it. The part of the capsule which lies beneath the tendon and behind the Y ligament has been slit, both to demonstrate its thickness, and to allow the head of the bone to rise as high as the ischiatic notch.

FIG. 15 shows the head of the bone dislocated below the tendon into the neighborhood of the sciatic notch. If the tendon were not present, the capsule would produce much the same effect in binding the head of the bone close upon the ilium without the interposition of the muscle.

9

socket by rotation outward, after the pulleys had sufficiently elongated or detached the Y ligament, especially if the pulleys were then relaxed and the thigh flexed a little. But as the obturator tendon and its subjacent capsule now lie between the head of the bone and the socket, they oppose the entrance of the head by a firm tendinous wall, which is drawn down as the head descends, and which no extension or rotation, short of its rupture, can displace or overcome.

The muscle is tense, and in its turn renders the ligament more tense, carrying the head of the bone backward and upward towards the ischiatic notch. The inversion, flexion, and adduction of the limb are thus augmented by the combined and reciprocal action of the ligament and the obturator muscle,— the latter being aided by the subjacent capsule, when that remains untorn.[1]

socket, exhibited much the same signs: the right leg was two inches shorter than the left; the knee and foot turned inward. An autopsy showed the posterior part of the acetabulum broken off, and the neck of the bone tightly embraced by the tendon of the obturator internus and the gemelli. (Cooper, Op. cit., Case LXXI., p. 113.)

[1] It has been before said, that, if the neck of the femur be further driven upwards, so as to rupture the obturator tendon and capsule, the luxation will become simply dorsal. Malgaigne correctly says, that "the ischiatic luxation leads frequently to the iliac luxation"; but he fails to identify the mechanism of the change, when he asks, "May not the former also be consecutive to the latter, in view of the fact that by flexion or strong traction the head of the bone may be drawn downwards from the dorsum to the notch?" (Op. cit., p. 831.) In the high dorsal dislocation, the small rotators would be so lacerated by the ascent of the bone, or by drawing it down to reduce it, when engaged in their interstices, that the luxation "on the ischiatic notch" would lose its distinctive features. In an interesting discussion upon a pathological specimen of hip luxation of five months' standing, where an unsuccessful attempt had been made to reduce the bone by longitudinal traction, M. Tillaux maintained, that, "in backward luxation of the hip, the capsule, and not the muscles"

REDUCTION.

The reduction is simple.[1] The head of the bone, having reached its present position by circumduction of the flexed limb inward, must be reduced by circumduction of the extended limb inward. When the thigh is raised perpendicularly to the floor, the head of the bone is unlocked and lies below the socket, and needs only to be jerked upward into its place; or the suspended pelvis may be depressed, or the thigh abducted and rotated outward, as in the common dorsal dislocation.[2]

(notably the obturator), "limits the movement of the head of the femur." (Société Impériale de Chirurgie, 1er Juillet, 1868, L'Union Méd., No. 79, p. 57.) It is true (see p. 66) that the obturator tendon and the posterior part of the capsule, which is next in strength to the Y ligament, mutually reinforce each other, so that, when the head of the bone rises behind the socket, it is generally engaged behind both these fibrous walls. But their mechanical action being identical, it is unnecessary to decide which under these circumstances would first be ruptured. The capsule yields fibre by fibre to the ascending bone, while the comparative elasticity of the tendinous muscle preserves it (Fig. 15), until, at the moment of the final rupture of the tendon, the dislocation has become practically iliac, and can now be reduced, though disadvantageously, with pulleys, by straight traction through the slit thus made behind the capsule. It may be safely asserted, first, that the tendon is usually present in these cases of the iliac luxation by inversion, unless the bone has risen so high upon the dorsum that the posterior capsule also has been ruptured, — and, secondly, that the tendon resists longest, and best characterizes the luxation.

[1] Mr. Nunneley, in the paper before quoted, expresses the contrary belief, that, in this luxation, reduction by manipulation will be more difficult, and will more frequently fail, than in any other form of dislocation to which the hip is liable.

[2] M. Lisfranc readily reduced a luxation "upon the sciatic notch" by the method of Desprès, twelve days after the accident. The pelvis being fixed, "he adducted the limb, at the same time flexing the thigh and leg; placing his fore-arm under the ham, and with his right hand grasping the ankle in order to use the leg as a lever, he

The laceration of the capsule is probably already sufficient, and will not need to be enlarged. It will be observed that by the flexion method this luxation and that upon the dorsum are reduced in the same way, and with equal facility.

I have had but two opportunities of satisfactorily identifying this dislocation in the living subject. The usual extended position of the luxated limb so endangers the obturator that its condition must often be a matter of uncertainty, although this luxation cannot be uncommon, compared with that higher up on the dorsum.

In the first case alluded to, which did not occur in my own practice, the patient, a middle-aged female, had fallen down stairs, and the limb had thus been subjected to a variety of forces. It was flexed, greatly inverted, and so advanced and adducted across the middle of the other thigh that I did not hesitate to recognize it, at sight, as a dislocation behind the obturator tendon: and yet it is possible that the bone may have been thrust between the rotators. With a view to its reduction, the limb was flexed, and a variety of movements were communicated to it, during which the bone slipped below the socket,—a change of position accompanied

instantaneously reduced the luxation by extension, outward rotation, and abduction."

For this case, which embraces the principles of flexion, abduction, outward rotation of the thigh, and the upward lift, see Observations sur Luxations, etc., M. Malespine, Arch. Gén. de Méd., Paris, 1839. See also Bull. de la Soc. Anat., 1835, p. 4, and 1836, pp. 45, 169.

Mr. Travers (London Med. Gaz., Nov. 22, 1828) relates a case of dislocation "upon the ischiatic notch," of six months' standing, which was reduced by pulleys, but in which the bone slipped out again while the thigh was flexed in bed during the night, — the obvious result of relaxation of the Y ligament. In subsequent unsuccessful efforts at reduction, the neck of the bone was fractured.

by a sharp report, probably due to the rupture of some fibrous band, or possibly the tendon of a rotator muscle.[1] It was afterwards lifted into its socket.

The following case admits of no doubt. While correcting these sheets, I was called to the Hospital to see a middle-aged man who had been struck upon the hip by a bale of hay three hours before. Having fallen over on his left side, the bale dropped from the story above, striking upon his right femur below the trochanter a little in front, dislocating it outward and downward. He said that two physicians had unsuccessfully tried, for an hour, with ether, to reduce it. He was in pain, sitting up in bed, the luxated thigh greatly inverted, and flexed so that it crossed the sound limb near the groin. (See Fig. 13.) After he was etherized and laid flat, the dislocated thigh, when drawn down, crossed the other at the junction of the middle and lower third, but still with great and firm inversion. This position of the bone, in connection with the facility of its reduction and the manner of the accident, indicated that the head, suspended at the trochanter by the Y ligament, was prevented from rising on the dorsum, so as to permit the descent of the knee, by some obstacle, which could be no other than the obturator tendon and the subjacent capsule, stretched across its neck,—also that the luxation was secondary, the bone having escaped below the socket before rising behind the tendon. After etherization, the knee came down somewhat, as the head rose behind the tendon. The hip was reduced by flexion, abduction, and eversion, with a slight upward jerk, at the first effort, and in three seconds from the moment the limb was grasped for flexion.

---

[1] In the dead subject the muscular fibres yield noiselessly.

Thyroid and Downward Dislocations.

1. Obliquely inward and downward on the thyroid foramen.
2. Obliquely inward and downward as far as the perinæum.
3. Vertically downward below the acetabulum.
4. Obliquely outward and downward as far as the tuberosity.

These dislocations, if we except that upon the thyroid foramen, are comparatively rare. In view of the frequency of accidents dislocating the bone while flexed or abducted, this rarity may be explained by the readiness with which the extreme downward luxations are converted into those upon the thyroid foramen or the dorsum.

Thyroid.

The bone escaping obliquely downward and inward beneath the socket, by a laceration of the inner side of the capsule, where it is thin and membranous, tends to follow the inclined plane of the pelvis towards the thyroid foramen, where it finds a lodgement.[1]

Signs.

The limb is unequivocally flexed and abducted, the

---

[1] For a case of thyroid dislocation occurring in a child six months old, see Lancet, May 16, 1868.

Figs. 16, 17, and 18. — Thyroid dislocation: Fig. 16 showing the front view, Fig. 17 the side view, and Fig. 18 the back view of the leg. The limb is seen advanced, abducted, and a little everted.

Fig. 19. — The mechanism of the thyroid dislocation, showing the Y ligament suspending the trochanters, while the head of the bone is lodged in the thyroid foramen, the trochanter resting on the acetabulum. (From a photograph taken in 1861.)

THYROID AND DOWNWARD DISLOCATIONS. 71

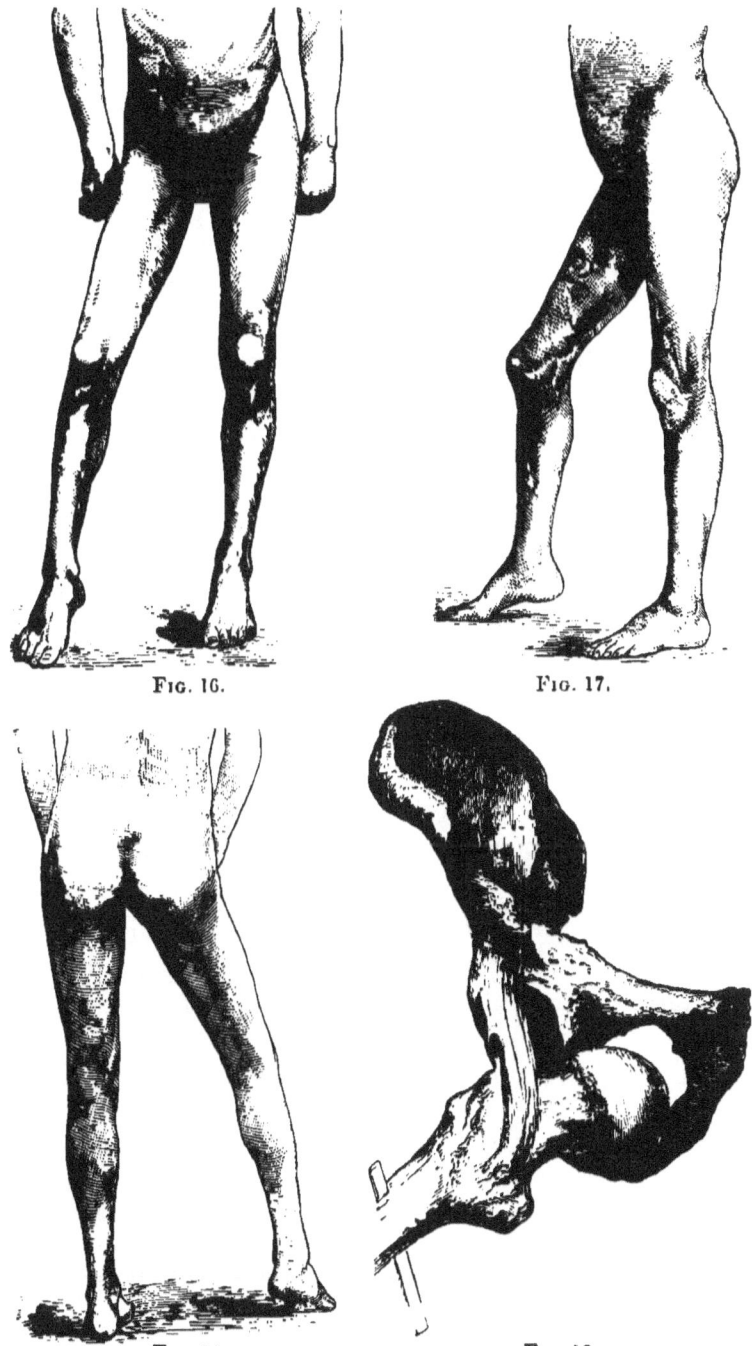

Fig. 16.   Fig. 17.

Fig. 18.   Fig. 19.

heel being raised from the floor, and the toe pointing outward and forward. The trochanters being arrested and suspended by the Y ligament, while the head of the bone descends from the socket, the thigh is flexed to an angle of about 35°, and also abducted, until the great trochanter, by swinging outward, gets a bearing on the acetabulum, and the adductor muscles become tense. The head of the femur likewise rests upon the pelvis, enabling the patient sometimes to walk tolerably well, and is hindered from rising towards the pubes, and even from re-entering the socket, by the inner margin of the acetabulum, the falciform edge of the lacerated capsule above, perhaps, contributing its resistance.[1]

The internal obturator muscle is not necessarily broken even in the complete dislocation. That part of the capsule which is attached near the ilio-pectineal eminence may assist the Y ligament in suspending the limb, the thigh becoming, in all cases, more flexed, when forcibly inverted. If the inner branch of the Y be ruptured, the bone is suspended by the great trochanter, and the eversion is diminished. Although it might be supposed that the extended psoas and iliacus muscles are concerned in the flexion, yet, after the Y ligament is divided, the tense fibres of these muscles produce a less degree either of flexion or eversion, and can be broken by depressing the knee. The long muscles of the anterior part of the thigh also become somewhat tense, and the head of the bone tends to escape towards the perinæum.

[1] In a case of M. J. Roux, the head of the bone had passed the thyroid foramen and reached the ischium; the leg was elongated, slightly flexed, and inclined outward. The thigh could be flexed, adducted, and abducted, but not extended. After unsuccessful traction, the luxation was reduced by flexion. (Revue Méd. Chir., Tom. IV. p. 364.)

VERTICAL DOWNWARD LUXATION.

Fig. 20.

Escaping directly downward, the head of the bone may remain upon the lower margin of the socket, the limb exhibiting less eversion than in the thyroid dislocation, but the luxation being practically of the same general character, provided the Y ligament be not ruptured. In Gurney's first case,[1] the eversion was slight; the flexion moderate, but, if carried beyond the sitting posture, painful; the knee lengthened by about an inch, standing and sitting; the foot capable of rotation inward and outward; and the limb able to support the weight of the body in walking, the patient having walked two hundred yards on the day of the accident, and a mile six days after. In a second, similar case, the patient could walk, the foot could be rotated outward and inward freely, but the limb could not be flexed to the sitting posture. The head of the bone was felt behind and below the acetabulum. In these cases, the bone obviously had a firm bearing below the acetabulum, which, while it was capable of supporting the weight of the body in walking, allowed rotation upon its convex surface. Flexion may have been hindered by the elongated extensors.

[1] See two interesting cases of dislocation of the thigh downward, by Edwin Gurney, Esq., Surgeon, Camborne, Cornwall. Lancet, 1845, Vol. III. p. 412.

Fig. 20. — Dislocation downward. The bone has descended towards the tuberosity, the flexion of the thigh being proportionate to the descent of the bone.

Hippocrates probably refers to a case of this sort, and not, as has been supposed, to luxation on the ischiatic notch, when he speaks of "the leg and foot appearing pretty straight, and not much inclined towards either side, ... the sole of the foot on its own line, and not inclined outward." Of the limb he says, "It becomes much shorter, and the patient can hardly reach the ground with the ball of his foot, and not even thus, unless he bend himself at the groins, and also bend with the other leg at the ham; or, if resting upon the foot, the hips protrude backwards far beyond the line of the foot." With a crutch, the patient "will walk indeed more erect, but will not be able to reach the ground with the foot; or, if he wish to rest upon the foot, he must take a shorter staff, and will require to bend the body at the groins."[1] This description indicates great flexion of the limb without inversion or eversion, and if it applies to a recent luxation, and not to an old one where the foot has been straightened by time, or unless we suppose the lesion here described to be the result of old hip disease, — an hypothesis which can hardly be considered possible, in view of the practical experience of the writer, — these signs are compatible only with dislocation beneath the socket.

Dislocations near the Tuberosity or Perinæum.[2]

When the thigh is thus strongly flexed, it is easy to imagine that the head of the femur, suspended by the Y ligament beneath the lower margin of the socket, pauses

[1] Hippocrates, Op. cit., art. 71.
[2] For a case of dislocation near the tuberosity, see Cooper, Op. cit., Case LXX. The limb was "considerably shortened and inverted," forming half a right angle with the body, — the shaft of the femur cross-

there, hesitating between the thyroid and the dorsal luxations. It has been found at various points in the interval between these luxations, and directed into the one or the other in attempts to reduce it. In extreme flexion, the head may reach as far as the tuberosity, on one side, and the ascending ramus of the ischium, and even the perinæum, on the other.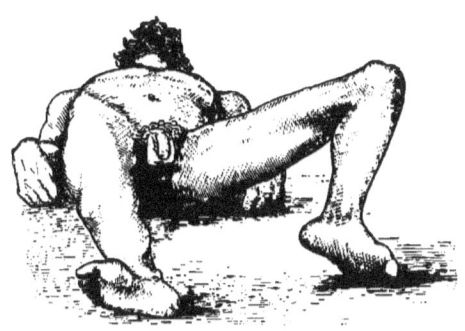

FIG. 21.

In short, in the dead subject, the ligament permits the head of the bone to descend, until the edge of its articular surface sweeps the centre of the tuberosity and the ascending ramus of the ischium. When found in these positions ing the symphysis, and being fixed there. At the autopsy the head was found on the tuberosity; the obturator internus was ruptured, and the ischium and ilio-pubic symphysis were broken, — complications which would not, however, necessarily modify the character of this dislocation.

For a case of perineal luxation, with autopsy, see Trans. London Path. Soc., Vol. X. p. 211. The thigh was much flexed and abducted, any attempt to adduct or depress it being met by resistance and pain. The head was felt in the perinæum. Reduction was effected by drawing the thigh vertically down from the pelvis, with lateral extension by a towel, aided by the knee of the operator in the groin. The capsular ligament was extensively detached, so that the head of the femur easily protruded. The "ilio-femoral ligament was detached at its outside, and partially separated from the neck of the femur, and a small rent extended from that point into the capsular ligament." Flexion here was doubtless due to the remaining inner band of the Y ligament.

FIG. 21.— Dislocation downward and inward towards the perinæum. As in the other regular downward luxations, the flexion is proportionate to the descent of the head of the bone.

in the living subject, so far as may be inferred from the reported cases, the ligament was not ruptured. Such being the position of the head of the femur in the dislocations with extreme flexion, the knee would occupy the extremity of the opposite spoke in an imaginary wheel of which the Y ligament should be the centre. The signs obviously vary with the position of the bone, the limb being always flexed[1] in proportion to the downward displacement of the head of the femur and the length of the ligament,— inverted when the femoral head is directed to the outside of the socket, and everted when it inclines to its inner aspect. If the head of the bone inclines a little to the inside, resting near the groove of the external obturator tendon, the limb is a little abducted, elongated, and rotated outwards, this being a first advance towards the thyroid foramen. If the head of the bone rests a little outside and behind the axis of the acetabulum, the rotation of the limb inclines proportionably inward, this being a step towards dislocation behind the tendon (Fig. 13), into which this luxation may be easily converted by depressing the knee. If the head of the bone is thrust down near the tuberosity, the limb is in extreme

[1] In some of the reported cases of downward dislocation where the head was felt near the tuberosity, it is impossible not to recognize the fact that the flexion of the thigh was less than it should have been, if the Y ligament was sound, as it was in the case of Stanski, for example. (See note, p. 78.) Such a case is that of Bouisson, where the head was on a level with the tuberosity, and the thigh said to have been but slightly flexed. (Gaz. Méd., 1853, p. 664.) The ligament may have been here ruptured in whole or in part, and if so, the dislocation was irregular. On the contrary, in a fatal case reported by Mr. Luke (Med. Times and Gaz., Vol. XVI. p. 12), the flexion of the leg is not alluded to, although the limb was lengthened one inch without inversion or eversion, and at the autopsy the head of the bone was found just below the acetabulum, and *the capsule was lacerated only below.* If so, the limb must have been flexed.

flexion, with, perhaps, adduction. If it is forced inward upon the perinæum, we naturally find, also, with extreme abduction, the thigh standing out at right angles with the body;[1] and as there is no firm bearing for the trochanter in the perinæum, as in the thyroid foramen, the toes may be inverted or everted.[2]

The obvious affinity and resemblance between these downward luxations, of which the thyroid is frequent and the others rare, need not be urged. The bone is suspended by the Y ligament, and when the head is displaced to one side of the socket the limb passes to

[1] The following case of perineal luxation is reported by Willard Parker, in the New York Journal of Medicine, March, 1852, p. 188. A man was standing beneath a canal boat, his legs apart, and received the weight of the falling boat upon his back. The left leg and thigh were found extended at a right angle with the body, and a little inverted, while the head of the femur could be felt in the perinæum behind the scrotum. Extension outward and downward carried the head of the bone into the thyroid foramen, whence it was reduced by carrying the femur across its fellow.

A similar accident happened to Pope's patient (Ibid., p. 198), upon whom a bank of earth fell while he was standing under it with his legs widely spread. The thigh was found to be at right angles with the body, inclined a little forward, the head of the bone projecting beneath the skin in the perinæum. Reduction was effected by lateral extension applied with pulleys to the upper part of the thigh, the leg being used as a lever.

The case of Amblard (quoted by Malgaigne, Op. cit., p. 876) was attended with a good deal of local pain, and with retention of urine. The lesion was caused by a fall from a cart upon the leg, which is supposed to have been already luxated by some twist before the patient reached the ground. The thigh was spread at a right angle to the body, with a little outward rotation. To reduce it, the leg was drawn downward and outward, and the head of the bone lifted outward with a towel. The head entered its socket by the way of the foramen ovale.

The case of Amblard showed eversion, while that of Parker exhibited inversion.

[2] The annexed wood-cut (Fig. 22) represents the specimen of Stanski

the other; or if the head is arrested directly beneath the cotyloid cavity, the limb is in simple flexion: the position of the limb thus indicating that of the head, which can then generally be felt, sometimes very distinctly, as in the perinæum.

In the downward dislocations, if the inner fasciculus of the Y ligament is ruptured, the head of the bone is inclined downwards by an inward rotation of the limb still suspended at the outer trochanter, — the head of the femur being then comparatively lower, and the limb less flexed, than if the inner fasciculus were unbroken. Such a state of the parts might exhibit the head of the bone in the neighborhood of the tuberosity without excessive flexion, but the limb would be greatly inverted.

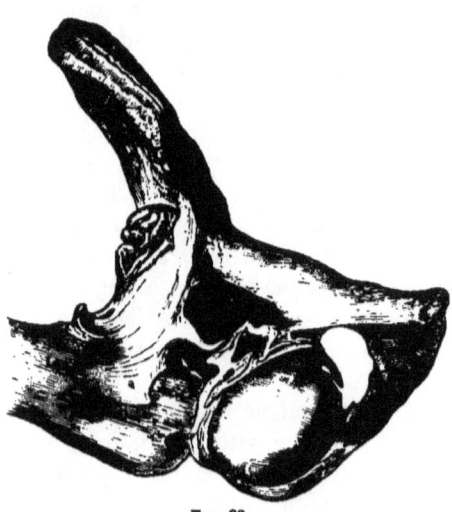

Fig. 22.

(Bulletin de la Soc. Anat., 1837, p. 296), and is taken from Malgaigne (Op. cit., Pl. XXVII., Figs. 4 and 5,) whose description is more complete than that of Stanski. It shows the anchylosed bones in a luxation of long standing, occasioned by the falling of a bank of earth upon a man while stooping. Although the dislocation is classed as thyroid by both these writers, the great flexion of the femur indicates that the head of the bone had passed downward and inward near the tuberosity, while the Y ligament remained entire: "a mass of bony stalactites, which seem to prolong the inferior iliac spine downward to join the internal face of the femur, to which they adhere to the extent of four centimètres, bending round its anterior face, and even behind it, to join the great trochanter" (Malgaigne), being, if we may judge from the figure, the tense and anchylosed Y ligament, beautifully illustrating this form of luxation. (Compare with Figs. 19 and

REDUCTION.

The thyroid dislocation is usually not difficult of reduction; but the following methods will illustrate the variety of expedients to which the surgeon may have recourse,—it being remembered, that the rent of the capsule, which is here thin, may be enlarged at discretion by circumduction of the flexed thigh inward.

1. *Rotation.* Flex the limb towards a perpendicular, and abduct it a little to disengage the head of the bone; then rotate the thigh strongly inward, adducting it, and carrying the knee to the floor. The trochanter is then fixed by the Y ligament and the obturator muscle, which serve as a fulcrum. While these are wound up and shortened by rotation, the descending knee pries the head upward and outward to the socket. As in reducing the pubic dislocation, the last half of this manoeuvre is an inward circumduction of the flexed limb accompanied with rotation, and is practically the

20). Or it may have been that in this case the external ligamentary band was broken, producing greater eversion and more flexion. I have examined only Malgaigne's lithograph of this specimen, in which the origin of the bony plate from the inferior spinous process is so clearly given, that, if correctly represented, there can be little doubt of its real character. Yet it is proper to say that M. Houel alludes to it as a part of the tendon of the psoas muscle. (Manuel d'Anatomie Pathologique, etc., par Ch. Houel, Professeur Agrégé, etc., Paris, 1862, p. 231.)

See also the case of Keate (London Med. Gaz., Vol. X. p. 19). The accident happened to a gentleman, who, while riding, fell into a ditch, his horse falling upon him and widely separating his legs. The limb was three and a half inches longer than its fellow, " much flexed," with very great abduction and eversion. The head of the bone was close to the tuberosity, and freely movable. It was reduced by the way of the foramen ovale,—the route of the luxation, as stated by the patient. This case may have been " irregular," because the operator was able to elongate or pull down the limb after reduction,—a circumstance which he attributed to a supposed fracture of the socket.

80   THYROID AND DOWNWARD DISLOCATIONS.

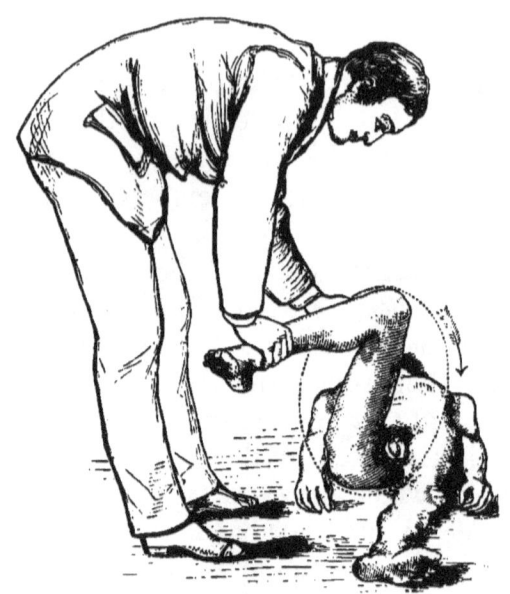

Fig. 23.

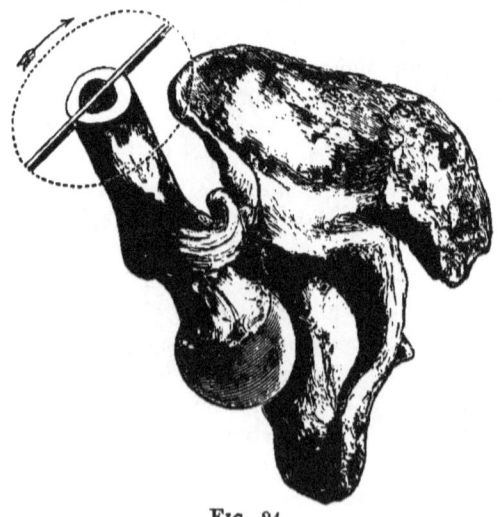

Fig. 24.

reverse of the flexion, abduction, and eversion, by which a dorsal dislocation is reduced from the opposite side of the socket. In this manœuvre, the action of the ligament may be aided, if necessary, by a towel passed round the head of the femur to draw it upward and outward. Rotation outward may be substituted for inward rotation.[1]

[1] In the paper already quoted, Dr. Markoe cites the two following cases of thyroid luxation reduced by rotation.

(Case 8.) Dr. Buck here reduced the bone by inward rotation, after two failures. In the third and successful attempt, the thigh was brought down from entire flexion to a little below a right angle, and again rotated inward, when the head of the bone slipped into its place.

(Case 9.) Markoe, in imitating Buck's method by rotation inward, unintentionally carried the head of the bone round the socket to the sciatic notch, from which position it was returned to the foramen ovale, and reduced by rotation *outward*, the knee being at the same time strongly adducted towards and behind its fellow.

These cases are instructive, as showing that the head of the bone is directed towards the socket when the Y ligament is wound upon the shaft by rotation, *whether inward or outward*, and they correspond to the results of my own experiments, made before I had read them. In the first case, the operator, in finally placing the thigh a little below a right angle, tightened the ligament, and directed the head upward, while at the same time a passage was left for the head of the bone between the trochanter and the socket. In the second, the surgeon, starting the limb at right angles, relaxed the ligament, engaged the head at its lowest point beneath the socket, and carried it by inward circumduction to the ischiatic notch. Had the thigh been now again placed in a vertical position, it could have been jerked up into the socket. It was, however, returned to the thyroid foramen, and reduced by outward rotation. In these cases, the head entered the socket while the knee was being depressed obliquely in-

Fig. 23. — The surgeon is here represented in the act of rotating and circumducting the flexed thigh inward.

Fig. 24. — The mechanism of the manœuvre shown in Fig. 23 is here seen. The inner branch of the Y ligament being wound round the neck, the head must rise towards the socket as the femur is depressed inward.

2. *Traction.* Flex the limb towards the abdomen, and draw the thigh outward by a towel passed round the upper part; or thrust it outward by applying the foot to the inside of the groin.[1]

3. Flex the thigh upward and outward, and drag or jerk it in that direction towards the socket. (Fig. 25.)

Fig. 25.

4. Lay the patient on his belly on the edge of a table, the injured thigh hanging, and the leg bent to relax the flexors; then draw the head of the femur outward with the aid of a towel.

5. Place him in a sitting posture, with a log, or post, or bedpost between his thighs, and pry the head outward over this fulcrum by means of the long shaft of the femur.[2]

ward. It may be superfluous to say, that, in Markoe's case, inward rotation would probably have reduced the bone, had the thigh been less flexed, or the manipulation been aided by oblique or vertical traction with a towel around the thigh at its upper part.

[1] In reducing a dislocation of this sort, flexion with lateral traction was successfully employed by M. Vertu. (P. A. Vertu, Thèse, No. 116, Arch. Gén. de Méd., 1836, p. 379.)

[2] In illustration of the flexion method, see Cooper, Case XLVI. Eight hours after a thyroid dislocation, attempts were made to reduce it by traction in the usual way, and were continued unsuccessfully until late at night, when, the pulleys breaking, further proceedings were deferred until the next day. The patient, having then taken two doses of tartar emetic, was carried into the operating theatre at 2 p. m.

Fig. 25. — Thyroid dislocation. Reduction by traction. The limb is flexed, abducted, and everted, relaxing completely the Y ligament. (From a photograph taken in 1861.)

6. Let him lie on a table, the limb flexed as usual. Then let an assistant, turning his back to the patient, carry the flexed knee over his own shoulder, grasping the foot, and endeavoring thus to lift the pelvis, while the surgeon draws the thigh outward by a towel in the groin.

7. Let the surgeon, facing the patient, place the flexed limb upon his shoulder, and, embracing the thigh near the pelvis, lift and direct the head of the bone towards the socket.[1]

8. Let the capsular orifice be enlarged by a little circumduction of the flexed thigh inward, as if to convert the thyroid into a dorsal luxation; and let the pelvis, suspended by the limb, be then depressed by the foot of the surgeon, while the thigh is drawn outward, if necessary, with a towel.

9. Convert the thyroid into a dorsal luxation, and proceed accordingly.

10. Most of these manœuvres may be executed while the patient lies on his sound side, if counter-extension be applied as a substitute for the weight of the body.

To reduce the other varieties of downward luxation, the femur should be flexed and its head drawn and guided towards the socket, during which manœuvre these dislocations are sometimes converted into that upon the thyroid foramen, or upon the dorsum below the tendon.

For the dislocation downward, we may employ verti-

Attempts at reduction were again made, and powerful extension employed for upwards of an hour without success. The tartar emetic was repeated in large doses, and the man, becoming faint, was placed in a *sitting posture*. Extension was then made, and after a short time the head of the bone slipped into the acetabulum.

[1] Method of Larrey, Malgaigne, Op. cit., pp. 853–855.

cal traction, rotating the femur a little inward to disengage the head; for the dislocation downward and outward, traction upward and inward, with abduction and rotation outward, if required to tilt the head; for the dislocation downward and inward, traction upward and outward. In these three injuries the femur is of course to be kept flexed, its head drawn and guided towards the socket by local pressure, or lifted with a towel, if necessary, with rotation outward, and abduction, when the bone is directly below or outside the socket, and with circumduction at discretion, when required to enlarge the capsular opening.

(See also the methods, 6, 7, and 8.)

Dislocation upon the Pubes, and below the Anterior Inferior Spine of the Ilium. (*Sub-spinous.*)

Dislocation upon the Pubes.

In this dislocation the head of the bone is felt upon the pubes; the limb is a little shortened and everted, abducted and advanced. A laceration of the inner aspect of the capsule allows the bone to escape obliquely upward, to a point upon the pubes distant in proportion to the violence of the force displacing it.[1]

Complete pubic dislocation is impossible until the capsule beneath the obturator internus is ruptured,[2] after which this muscle everts the limb until the tro-

---

[1] Larrey is said to have seen a case of pubic dislocation in which the femur was flexed at nearly a right angle with the body. (Hamilton, Op. cit., p. 655.) It is fair to suppose that it could have been brought down to the usual position.

[2] In an autopsy of a case of pubic dislocation, in a paper by Mr. Bransby Cooper (Guy's Hospital Reports, 1836, Vol. I. p. 82), the gemini and quadratus femoris had suffered from laceration and subsequent ulceration, implicating all the outward rotators of the thigh.

DISLOCATION ON THE PUBES. 85

chanter bears upon the pelvis. If this muscle is ruptured, the psoas and iliacus, binding the neck of the bone to the pubes, may produce a degree of eversion; but the principal agent of eversion even then is the Y ligament, which also embraces the neck. The untorn

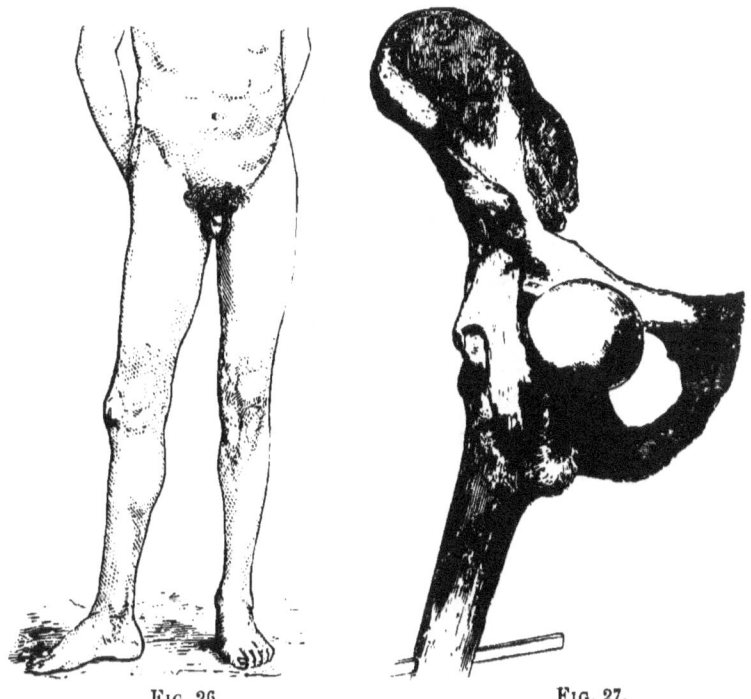

Fig. 26. Fig. 27.

capsular fibres and the obturator muscle are agents in preventing flexion, their insertions being lower than the head of the displaced bone, which then becomes a fulcrum, the lever in flexion being the shaft; but in pubic dislocation nearer to the iliac spine, the obturator is not tense, and flexion is then probably hindered by

---

Fig. 26. — Pubic dislocation. The foot is everted, the thigh advanced and abducted.
Fig. 27. — Pubic dislocation. The head of the bone is seen in the groin, suspended by the Y ligament. (From a photograph made in 1861.)

the outer and inferior parts of the capsule, when they still exist. Both the muscle and the capsule act in preventing inversion. Dislocation to the neighborhood of the symphysis implies a rupture of the inner branch of the Y ligament.[1]

Dislocation below the Anterior Inferior Spine of the Ilium, or Sub-spinous.

The head of the bone ranges along the pubes, displaced according to the violence and direction of the injury. If thrust directly upward, the bone may lie beneath the Y ligament and the inferior iliac spine; but this displacement requires that the upper part of the capsule should be completely detached from the edge of the socket. The firm bearing of the neck against the Y ligament may then explain how the patient has in some recorded cases been able to walk immediately after this accident.[2] The limb is still everted, but less abducted or advanced, and the head of the bone is plainly felt in its new position,—in the

---

[1] A careful autopsy of pubic luxation is recorded in a communication of M. Aubry, read by M. Maisonneuve, to the Société de Chirurgie (Arch. Gén. de Méd., Paris, 1853, p. 355). The head of the bone projected in the groin, the limb was rotated outward with flexion, a little abduction, and shortening to the extent of one quarter of an inch. The autopsy showed the psoas and the crural nerve upon the anterior surface of the neck. Half the anterior circumference of the capsule was torn at a quarter of an inch from its cotyloid insertion, the neck of the femur being held in a sort of button-hole between its fibrous edge and the cotyloid rim. Flexion of the thigh obviously relaxed this fibrous band, liberating the neck; extension produced the contrary effect, strangulating the neck. Of the muscles, the external obturator was relaxed; the pyriformis, internal obturator, and gemelli appeared elongated.

[2] See Malgaigne, Op. cit., pp. 844, 845.

absence of which evidence, the shortening and eversion might possibly be mistaken for fracture of the neck.[1]

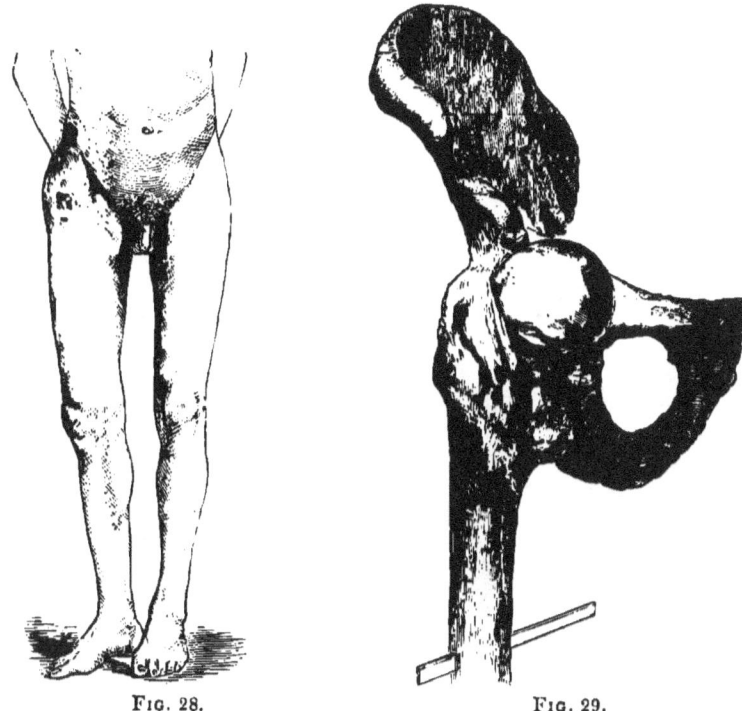

Fig. 28.   Fig. 29.

When the bone has been thus displaced, the psoas and iliacus tendon is sometimes thrown off the neck of the femur, towards the pubes, where it then lies slack. But even when in place, the action of this tendon is wholly secondary to that of the Y ligament

[1] The annexed figure (Fig. 30) from Malgaigne (Op. cit., Pl. XXVII.,

---

Fig. 28. — Pubic dislocation nearer the spine. The limb is here seen everted, but is usually a little more advanced and abducted. Nélaton, however, describes a similar absence of flexion. (Clinical Lectures on Surgery by M. Nélaton, from Notes taken by W. F. Atlee, M. D., Phila., 1855, p. 213.)

Fig. 29. — Sub-spinous dislocation. The neck of the bone is seen lying beneath the Y ligament, which is tightly stretched across it. (From a photograph made in 1861.)

in producing either flexion or eversion, as may be shown by its division, after which the position of the dislocated bone remains unchanged, — while, if the Y ligament be divided without the tendon, the bone drops to a position near the thyroid foramen, with little flexion : an attitude of the limb resembling the irregular dislocation towards the perinæum or on the tuberosity.[1]

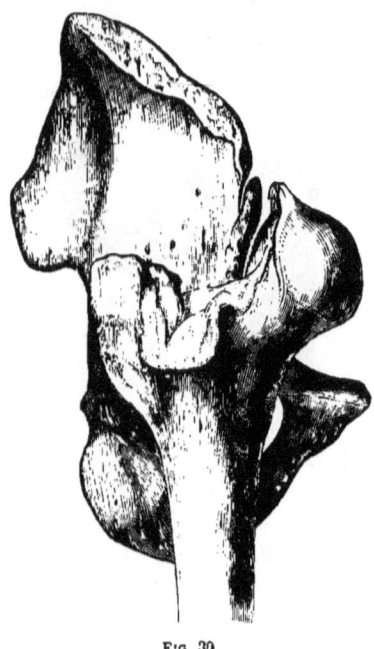

Fig. 1) represents a specimen elaborately described by M. Gely (Bulletin de la Soc. Anat., 1840, p. 303). The accident occurred to an insane person, a long time before death. The neck of the bone rides upon the inferior spine, and the leg is much everted. This eversion may, perhaps, be referred to pathological changes, but may have occurred at the time of the accident. Gely rightly supposes the weight of the body to have been supported by the upper part of the capsule, reinforced by the tendon of the rectus muscle. (Op. cit., pp. 320, 327.)

[1] If the head of the bone be still further displaced outward, it lies beneath the inferior spinous process, as in the case of Wormald (London Med. Gaz., Jan., 1837, p. 164), where the limb being somewhat everted, abducted a little, and shortened half an inch, the new cavity was formed in part by the upper portion of the cotyloid ligament. The patient, who died twenty-six years after the accident, was said to be able to walk well, being " engaged in carrying out beer for a publican in Portugal Street," — a statement which Malgaigne oddly translates, "*pour porter un mort au cimetière.*" (Op. cit., p. 871.)

Fig. 30. — Sub-spinous dislocation.

DISLOCATION ON THE PUBES.   89

REDUCTION.

I have never met with pubic dislocation in the living subject, and am therefore unable to speak of the extent of a difficulty in flexion alluded to by some writers as characteristic of this luxation. But there is ample evidence that this difficulty is neither insuperable nor constant. The pubic dislocation has often been reduced by flexing the limb; and if the obturator tendon and its subjacent capsule resist flexion in the living as in the dead subject, the limb needs only to be drawn down towards the socket while in the act of being flexed.

If the bone has been thrust upward between the Y ligament and the pubo-femoral band, and the capsular orifice be small, this band may be ruptured by circumduction or even rotation of the flexed thigh inward. But well-marked pubic dislocation usually implies a rupture of the capsule which extends to its inner and lower aspects.

It is difficult to reduce the pubic dislocation by straight extension, and various accidents have happened in attempting it.

The reduction may be accomplished in a variety of ways, among which are the following, combining angular traction and rotation.

1. *By Traction and Rotation.*— Flex the limb to a right angle, while drawing it down; rotate either inward or outward, and, directing the head of the bone by its shaft, rock it downward into its place.[1]

---

[1] Two cases of pubic dislocation skilfully reduced by manipulation are reported by Dr. E. J. Fountain, of Davenport, Iowa, in the New York Journal of Medicine, etc., January, 1856, p. 69. In the first case, the patient was laid upon the floor on a quilt, made insensible with chloroform, and the limb was rotated outward. The leg was then

2. While extending the limb horizontally, with counter-extension by the foot in the perinæum, raise the patient to a sitting posture, counter-extend against the pubes, and rotate inward.

3. The same method may be pursued, the patient lying on his belly on the edge of a table, or on his sound side.

4. (See Reduction of the Thyroid Dislocation, Nos. 7 and 8.)

flexed and carried across the opposite knee and thigh, the heel kept well up and the knee pressed down. This motion was continued by carrying the thigh over the sound one as high as the upper part of the middle third, the foot being kept firmly elevated; then the limb was carried directly upward by raising the knee, which was gently oscillated, when the head of the bone dropped into its socket. The time of this operation was from twenty to thirty seconds, and the force slight. In a second case, rotation and flexion produced greater pain, and the limb was less movable. Here also the knee and foot were rotated outward, the leg then flexed across the sound thigh, the heel kept up and the knee pressed down. The whole was carried in this position across the sound thigh directly upward to the flexed position, the operator holding the foot firmly up and making oscillations with the knee, when the head of the bone slipped into the socket. About twenty seconds sufficed for the operation, which was performed without the use of chloroform.

It will be observed in these cases that no real difficulty was encountered in flexion. The limb was flexed, and the vertical femur, rotated outward, was rocked down into its place. The outward rotation of the flexed femur made the outer branch of the Y ligament tense, with an interval, through which the head of the bone, already rotated to a point just above the socket, descended into it. Perhaps, as Dr. Fountain recommends, the whole manœuvre should be commenced with an outward rotation, to be maintained till the reduction is accomplished; but it seems to me that this rotation is unnecessary, until after the limb is flexed.

Devilliers and Aubry each reduced a pubic dislocation by flexion and rotation inward instead of outward, and Larrey by simple downward pressure at the groin, with the knee over his shoulder. (Malgaigne, Op. cit., pp. 853, 854.)

DISLOCATION ON THE PUBES. 91

5. Flex and abduct the limb and draw it outward, at the same time pressing the head downward and outward.

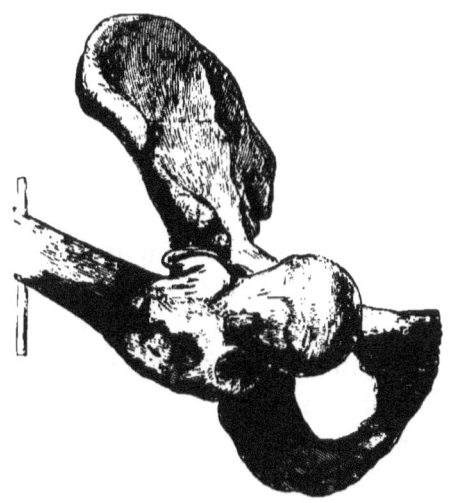

FIG. 31.

By Rotation. — Reduction by rotation is to be accomplished by much the same method as in the thyroid dislocation, except that in the pubic luxation the flexed limb should be carried across the sound thigh at a higher point. First, semi-flex the thigh, to relax the Y ligament, at the same time drawing the head of the bone down from the pubes. Then semi-abduct and rotate inward, to disengage the bone completely. Lastly, while rotating inward and still drawing on the thigh, carry the knee inward and downward to its place by the side of its fellow. As in the thyroid luxation, this manœuvre guides the head of the bone to its socket by a rotation which winds up

---

FIG. 31.—Pubic dislocation. Reduction by traction. The limb has been here flexed and abducted, for reduction by traction and local pressure: the abduction is represented as greater than necessary. (From a photograph made in 1861.)

and shortens the ligament, enabling the operator, by depressing the knee, to pry the head of the bone into its place.

Briefly, while drawing upon the thigh, flex and abduct it, to disengage the head; then rotate inward, and, when the bone leaves the pubes, continue the rotation while straightening the limb; or circumduct the bent limb inward.[1]

Aid these manœuvres by drawing the flexed groin outward with a towel, or otherwise depressing it.[2]

If by these combined movements of traction, leverage, and rotation — of which the Y ligament and the obturator tendon, when it is unbroken, are the centre — the luxation is not reduced, it will perhaps be converted into one near the thyroid foramen, the rules for the reduction of which will then apply here.[3]

Anterior Oblique Dislocation.

The remaining luxations imply a free laceration of the tissues about the joint, and sometimes of a part of the Y ligament itself.

[1] See case of Dr. J. M. Irvine (Brit. Amer. Journal, March, 1861, p. 282). A complete pubic dislocation of the right hip was reduced by flexing the thigh upon the pelvis, carrying the knee over the umbilicus to the left side of the body, and thence to a state of extension, when the head slipped in.

[2] Baron Larrey has reported a case of dislocation in front of the horizontal portion of the pubes, which he reduced by suddenly raising with his shoulder the lower extremity of the femur, while with both hands he pressed the head of the bone downward. (Hamilton, Op. cit., p. 657, and Lond. Med. Chir. Rev., Dec., 1820, p. 500.)

[3] Mr. Annandale, after some unsuccessful manipulation, succeeded by flexion in reducing a pubic dislocation of three days' standing, but used pulleys to withdraw by outward extension the head of the bone from the pubes. (Thos. Annandale, F. R. S. E., etc., Asst. Surg. Royal Infirm., Edinburgh Med. Jour., 1867, p. 997.)

ANTERIOR OBLIQUE DISLOCATION. 93

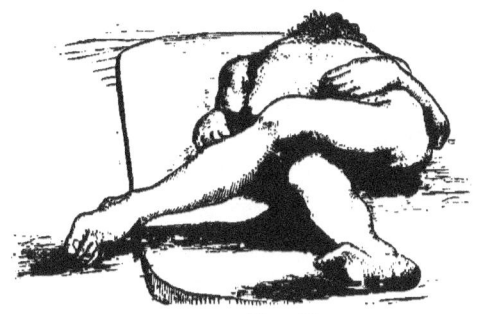

Fig 32.

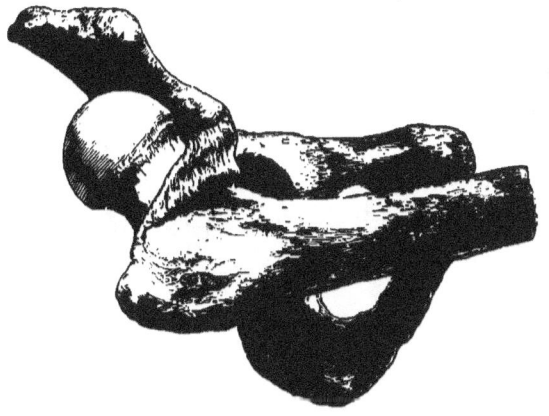

Fig. 33.

Figs. 32, 33, 34. — Anterior oblique dislocation. The limb is here extremely everted, crossing the other above the knee. The general anatomical character of the luxation is seen in Fig. 33, where the Y ligament is still entire, the limb crossing the other high up. As the limb descends towards a perpendicular, the outer fibres of the ligament yield, until, as it reaches the position seen in Fig. 37, only the inner fasciculus remains. The head of the bone is then hooked over this inner fasciculus, as seen in the dotted line (Fig. 37), and the supra-spinous luxation is complete. If now thrust back upon the dorsum, the dislocation is simply the everted dorsal, as shown in Fig. 40, where, however, the toes may be inverted at will.

Fig. 33. — Anterior oblique luxation. By depressing the shaft of the femur the head rises over the inferior spinous process, as the external part of the ligament yields.

ANTERIOR OBLIQUE DISLOCATION.

In a common dorsal dislocation, let the leg be carried across the symphysis, so that the outer and convex surface of the socket shall correspond to the hollow beneath the neck of the femur. With some force the thigh can now be everted, and afterwards brought down across the upper part of its fellow. It is here firmly locked, with great shortening and some eversion, the limb facing forward and obliquely crossing the opposite thigh, while the toe points outward, — a position not wholly ungraceful, and suggesting some attitudes in dancing. (Figs. 32 and 33.)[1]

If, in this position, it is desired to bring the limb towards a perpendicular, the outer branch of the Y ligament must be ruptured. Thus liberated, it hangs suspended by the inner ligament, and becomes capable of lateral motion and of rotation; and this is probably the condition under which supra-spinous luxation, although rare, usually occurs. (Fig. 35.)

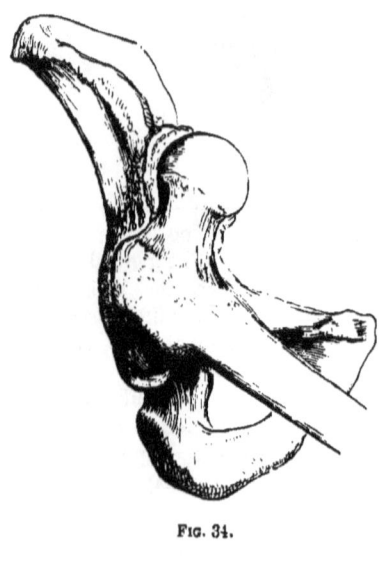

Fig. 34.

[1] For a description of the annexed wood-cut, taken from Cooper, and which exhibits the position of the anterior oblique luxation, see case of Oldknow (Guy's Hosp. Rep., No. 1, p. 97), also Cooper (Op. cit., Case LXVII.). The foot is said to have been very much everted, only the toes touching the ground. But the patient had lived twelve years after the accident, and something may be allowed for pathological changes. For a larger figure representing this dislocation, see a paper of Bransby Cooper, Guy's Hospital Reports, 1836, Vol. I. p. 81.

Fig. 34. — Anterior oblique luxation.

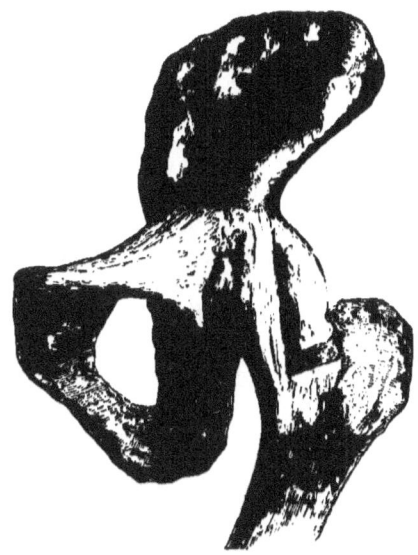

Fig. 35.

The anterior oblique dislocation may be reduced by inward circumduction of the extended limb across the symphysis, with a little eversion, if necessary, to disengage the head of the bone. Inward rotation then converts this into the common luxation upon the dorsum.

DISLOCATIONS IN WHICH THE OUTER BRANCH OF THE Y LIGAMENT IS BROKEN.

SUPRA-SPINOUS DISLOCATION.[1]

The head of the bone has been found above the inferior spinous process, the neck lying across the edge

[1] See case of Cummins (Guy's Hosp. Rep., Vol. III.). Cooper (Op. cit., Case LXV.) cites this case as anomalous, illustrating it with a

FIG. 35. — This figure is intended to show, in diagram, the external portion of the Y ligament detached, as in the supra-spinous and everted dorsal luxations.

of the pelvis, the trochanter turned back, and, as is said, not readily discovered. The limb was shortened two or three inches, a little abducted, and everted, — this eversion being sometimes so great that the toes pointed backward, although, in one of the cases related by

figure which represents the head of the bone as projecting farther upon the abdomen than the context indicates. The leg was shortened three inches, and could not be drawn down. The limb, which was much everted, could not be rotated inward. Cooper considers this to be "a variety of dislocation hitherto unknown."

Travers (Med. Chir. Trans., Vol. XX. p. 113) thus describes a case. "The trochanter is felt below and to the outer side of the anterior superior spinous process of the ilium. The neck of the bone lies apparently between the two anterior spinous processes, so that, when the patient is erect, the limb seems as it were slung or suspended from this point."

Sir Astley Cooper (Case LXII.) cites a case of old dislocation "on the pubes." An accurate account of the autopsy, with the annexed figure, is given by N. Cadge, F. R. C. S., Norwich (Med. Chir. Trans., Vol. XXXVIII. p. 88). The left leg was full an inch and a half shorter than the right; the toes were turned outward; and while the body lay on its back, the foot rested completely on the outer border. A large, globular, bony tumor was felt in the groin, close to the superior spine of the ilium. On dissection, the head of the femur was found in the interval between the anterior superior and anterior inferior spinous processes of the ilium. The head of the femur was covered with a complete bony cap, lined with a dense, pearly-white tissue, resembling

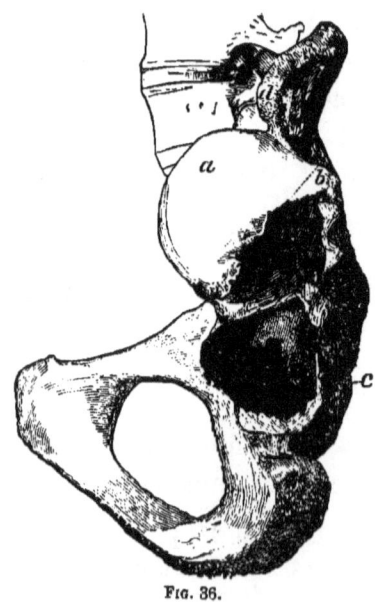

Fig. 36.

Fig. 36. — Supra-spinous dislocation. *a*, bony cap; *b*, fractured margin of ditto; *c*, socket; *d*, superior spinous process of ilium.

Cooper, they could be brought forward again to the side of the other foot. Another important feature was, that the shortened limb could not be drawn down.

In this luxation, the neck was doubtless hooked over the Y, and perhaps over the tendon of the rectus muscle also; so that direct extension, short of the rupture of this ligament, was worse than useless. The head of the bone had been thrust above and outside the Y ligament, upon which, in its return, the neck of the femur had engaged itself, the main branch of the Y then lying behind the neck, and so wound around it as to produce great shortening.

In the supra-spinous luxations, eversion is due to the internal obturator, when it remains entire, but also to the tense ligament.

The muscles inserted into the back of the trochanter, especially the obturator internus, hinder the head of the bone from advancing upon the spinous process; but when they are divided, the head advances towards the abdomen. The first degree of supra-spinous luxation, which is represented in the annexed wood-cut (Fig. 37), requires the rupture of only the outer fibres of the Y ligament, and is but a slight exaggeration of the anterior oblique luxation (Fig. 33). But when the bone projects fairly upon the abdomen (as illustrated by the dotted line, Fig. 37), only the inner fasciculus remains.

It may be remarked, that the anterior oblique dislocation, while it is also supra-spinous, differs from it in the comparative soundness of the ligament, which

---

fibro-cartilage. (Fig. 36.) The edge of the new cavity was connected with the neck of the thigh-bone by a thick capsular ligament. The rectus muscle, which had been torn from its origin, was inserted into the edge of the new cavity, — a condition that suggests the ascent of the bone above the inferior spinous process of the ilium at the time of the injury, with rupture of the Y ligament. This luxation may have been supra-spinous or irregular.

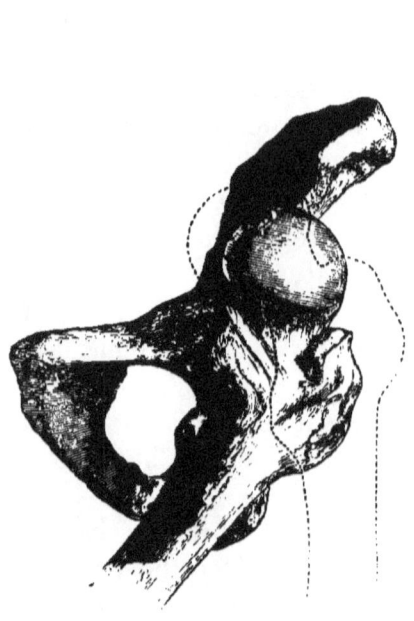
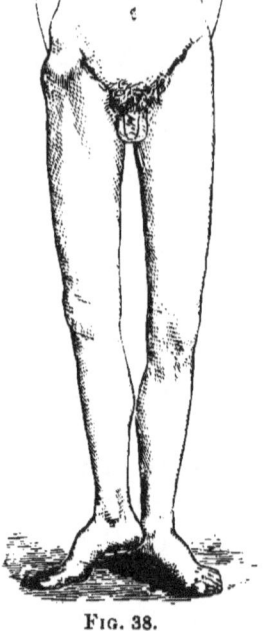

Fig. 37.   Fig. 38.

Figs. 37 and 38. Supra-spinous dislocation. (See note, page 93.)

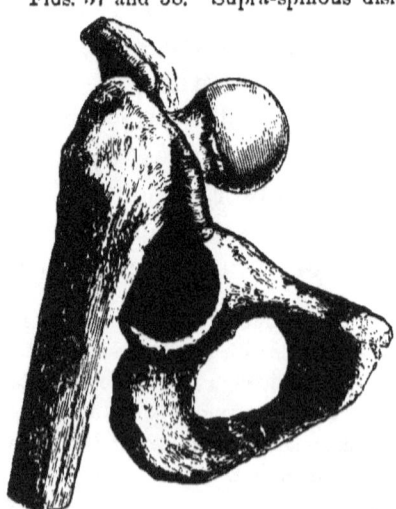

Fig 39.

Fig. 39. — This wood-cut is intended to show a possible posterior oblique luxation, the Y ligament being entire, the head of the bone thrust across it, and the shaft locked behind the tuberosity. In the dissected bones, it will be found that the femur is firmly locked, the limb being directed backward, and the foot somewhat everted. But there is no authority to show that such a position of the leg has been found in the living subject. By forcibly advancing the knee, the outer branch of the ligament is ruptured, and the luxation becomes supra-spinous.

compels the limb to assume an oblique position. In the latter luxation, the outer band is broken, the limb is more movable, and the term "supra-spinous" has been reserved for this, as probably the more common of these two rare varieties.

REDUCTION.

After extension by pulleys in the axis of the body has failed, reduction of this luxation has been accomplished by extension downward and outward, with some manipulation of the head of the bone, and probably with rupture of the ligament. It is obviously a better plan to unhook the neck by circumduction of the extended limb inward, with eversion enough to disengage it from the edge of the pelvis. The head then lies upon the dorsum, and, if the outer branch of the Y is broken, is not inverted. The reduction may then be accomplished as usual in the dorsal dislocation.—although rotation would be less effectual than if the ligament were entire.[1]

---

[1] The following case well illustrates the mechanism of the supra-spinous luxation, and is taken from Dr. Hamilton's work on Fractures and Dislocations (Op. cit., p. 649). " Lente relates a case [of ischiatic luxation] under the care of Dr. Hoffman, in the New York City Hospital, in which, when the extension was suddenly relaxed by cutting the cord, and the thigh at the same instant was abducted and rotated outwards, the head of the femur left the ischiatic notch and rose upon the dorsum ilii, assuming a position directly above the acetabulum and below the anterior superior spinous process, from which position it was with great difficulty subsequently returned to the socket."

If this luxation was really "ischiatic," as stated, and therefore " below the tendon," the forcible outward rotation of the thigh ruptured both the tendon and the outer part of the Y ligament, or in any case the latter, after which the head of the bone was free to turn forward

Everted Dorsal Dislocation.

It has been before stated, that inversion of the limb, in the dorsal luxations, is due to the tense outer branch of the Y ligament. When the injury has been such as to rupture these fibres, the limb may still be inverted, but it can also be freely everted. Having escaped from the socket under these circumstances, the bone may occupy any point upon the dorsum within the range of the inner fasciculus. The limb is then shortened in proportion to its upward displacement, the foot being sometimes everted a little, sometimes lying flat upon the bed, or even directed backward, the head of the femur facing accordingly, and, as has been elsewhere remarked, in the direction of its internal condyle. The femur is suspended midway between the inner branch of the Y and the obturator tendon. Theoretically, it may be luxated either below or above this tendon; but in the former case, the degree and nature of the force required to break the outer band would be likely to rupture the tendon also. If the head of the femur is driven upward and backward above the obturator tendon, the same forced eversion which would

Fig. 40.

and rise on the ilium toward the spine, the limb being of course everted, and the head of the bone perhaps engaged above the remaining ligament.

Fig. 40. Everted dorsal dislocation. (See note, p. 93.)

sever the inner branch of the Y ligament would relax this tendon, and so contribute to prevent its rupture. The tendon may then lend its aid in giving position to the limb.[1]

[1] For an old case of this sort, with an analysis of the muscular action, see a paper by Dr. Gordon, Dublin Hosp. Gaz., Nov. 1, 1845, p. 87.

Mr. G. R. Symes has described a case (On an Unusual Form of Dislocation of the Hip Joint, by Glasscut R. Symes, one of the Surgeons of Stevens's Hospital, Dublin Quar. Jour. of Med. Science, 1864, Vol. XXXVIII.) in which the right leg was shortened two inches, the foot extremely everted, the buttock flattened, the head of the femur two inches below the anterior superior spinous process of the ilium. The limb remained unreduced after protracted efforts by manipulation and pulleys, during which it was repeatedly inverted and everted. The failure to reduce the limb was attributed by Mr. Symes to a "buttonhole" laceration. In a similar case, or even if the head of the femur were engaged in the interstices of the rotators, I should attempt to liberate it by circumducting it to the thyroid foramen.

A case of everted dorsal dislocation has been reported by Dr. Van Buren (Contributions to Practical Surgery, by W. H. Van Buren, M. D., etc., Phila., 1865, p. 157). The limb was shortened an inch, and slightly everted, there being some obstacle to inversion. The trochanter was an inch and a half behind and above its usual position, and the head of the bone was obscurely felt in the back part of the sciatic notch. After repeated attempts at reduction by manipulation, the bone was reduced by pulleys applied to the thigh in a flexed position.

For a case probably everted dorsal, but classed by Cooper as anomalous, see Morgan, Guy's Hosp. Reports, No. 1, p. 82. The left leg was shortened two inches, the foot excessively everted, so as almost to give the toes a direction backward, but, when placed side by side with the other foot, remained in that position. The leg was to some extent susceptible of all the natural motions, with the exception of rotation. The trochanter could not be felt, but the head of the bone was apparently lying between the anterior inferior spinous process of the ilium and the junction of that bone with the pubes. Traction was made from the knee against counter-extension with the foot in the perinæum. The patient was then directed to raise his shoulders from the bed, extension was suddenly increased with forcible inward rotation of the thigh, and the head snapped into the socket.

Reduction.

The limb should be flexed and inverted, with adduction, if necessary, to make room for the head of the bone to slide upon the ilium, and the dislocation is then practically a simple dorsal dislocation, and easily reduced; or if not, perhaps the whole upper part of the capsule is detached, making the luxation irregular.

The rupture of the outer fasciculus of the Y ligament deprives the operator of much of the advantage of rotation. The limb, after flexion and rotation inward, may be reduced by direct traction towards the socket, with local guidance.[1]

[1] The following interesting case (reported by Dr. Shrady, in the New York Journal of Medicine, March, 1860, p. 255) occurred in the hospital wards of Dr. Willard Parker. The patient was crushed to the ground by a gravel car falling upon the small of his back. The left limb was rotated outward and shortened three inches, the thigh slightly adducted and flexed, the knee slightly advanced and semi-flexed, and the toe so everted that the heel rested against the inner aspect of the opposite leg, just above the ankle; passive rotation was very painful, the buttock of the affected side was much fuller than the other, and the post-trochanteric depression was obliterated. Only the tips of the toes touched the floor. The vertical distance from the trochanter to the crest of the ilium was shortened three quarters of an inch. (If this statement is correct, the apparent shortening of three inches was probably due to the flexed knee.) The head of the bone could be felt, but not very distinctly, in a direction forward and upward from the trochanter. Several efforts to reduce the limb by flexion and adduction were unsuccessful. The thigh was at last rotated inward, extension made in the direction of the socket, and the head of the bone guided by direct manipulation into its place.

IRREGULAR DISLOCATIONS.

(In which the Y ligament is wholly broken.)

IN rare instances the Y ligament may be completely ruptured by forced extension of the limb, or by an upward thrust, while the lower half of the capsule remains comparatively sound. But it has been shown that the position of the great majority of dislocations is determined by this ligament; and until it is likewise shown, that, when it is broken, the luxated limb will be compelled, in obedience to other mechanical agents, muscular or capsular, to assume positions equally constant, it is fair to consider such luxations as irregular. When any mechanism shall be shown always to give to a luxated limb, after the Y ligament has been torn asunder, the same position under the same circumstances, the luxation may be withdrawn from the present category, and classed as "regular."

When the Y ligament is wholly broken, and the head of the femur is dislocated upward upon the edge of the socket, either inside or outside the iliacus tendon, there is little or no shortening, and no flexion; but the eversion of the foot is marked. The head is felt in the groin, and is reduced by flexion and inversion.

If the head of the bone, under these circumstances, be displaced towards the thyroid opening, there is abduction of the leg, produced chiefly by the fascia lata, with some flexion, due to the adductors; but the flexion is less than in the regular thyroid dislocation, and the knee can be depressed, with a little effort, to the natural position. It is possible that such a dislocation might simulate the thyroid displacement; but it may be distinguished from this by the greater abduction and less considerable flexion of the limb.

If the head be now carried farther downward, the flexion becomes more considerable, though less than if the Y ligament were entire. Such may have been the condition of the parts in some of the cases of downward dislocation before referred to, where the head was said to have been felt near the tuberosity, and where the flexion was inconsiderable.

If the head of the bone be now carried behind the tendon of the obturator internus muscle, there is a flexion of the femur at an angle of 45°, but with such exaggerated inversion as to distinguish it from the regular dislocation below the tendon. The thigh then faces completely inward, and, instead of crossing its fellow, is even a little abducted. The leg, which is bent by the tense flexors of the thigh, stands at right angles with it.

If the head of the bone be carried upward upon the dorsum, the limb, while it faces directly inward towards its fellow, is no longer flexed, as in the regular dorsal dislocation, but lies flat upon the table. The head, being now detached from the socket, may be carried round upon the dorsum and hooked above the rectus muscle in front, — a position of the parts which, owing to the great strength of the Y ligament, is probably less frequent than the regular supra-spinous and everted dorsal luxations, where a portion of this ligament still remains.

The Y ligament being destroyed, an upward and backward dislocation, if attended with accidental inversion, may be held in that position by the lower part of the capsule, which, however, is readily ruptured by outward rotation or circumduction.

IRREGULAR UPWARD LUXATION.

The bone may be thrust upward upon the inferior spine or above it, with rupture of the Y, but can then be drawn down as far as the remaining capsule will allow, unless detained by being hooked over the muscles arising from that point.[1]

IRREGULAR DOWNWARD LUXATION.[2]

This variety should be distinguished from that in which the Y remains entire, — described in connection

[1] For several cases, of which the description is incomplete, but in which the limb was rotated outward, the head of the femur being outside the anterior inferior spinous process of the ilium, see Malgaigne, Op. cit., p. 869. In a patient at St. George's Hospital, the head of the bone was dislocated upward upon the inferior spine of the ilium, and a little to the outside, the upper half of the capsule being largely torn. (Lancet, 1840–41, Vol. II. p. 281.) In Gerdy's case, reported by Baron, the upper half of the capsule was torn, but the round ligament was only half broken. In this case the limb was reduced by flexion, the head of the bone being pressed towards the socket. (Malgaigne, p. 870.)

See also the case of Adam Hunter, Edinb. Med. Chir. Trans., 1824, p. 171. The limb was shortened one inch, and the toes turned inward. The head of the bone was over the sciatic notch, — the gluteus minimus, pyriformis, obturator internus, and other small muscles, being ruptured. The capsule was entirely detached from the femur, so that, when the ilio-femoral muscles were divided, the limb was separated from the trunk. The head was said to have been bound down firmly on the sacro-sciatic notch by the gluteus medius, which passed over the neck of the bone. In the absence of the capsule, it is quite possible that the gluteus medius, beneath which the head of the bone was found, together with the anterior flexors of the thigh, exercised a certain controlling influence on the position of the limb; and yet, after dividing the whole capsule in a recent subject, and engaging the head of the femur fairly under the gluteus medius muscle, I have found that rotation ruptured its fibres with little effort.

[2] The case of Keate (see p. 79) may have been irregular, be-

with the thyroid luxation. In the latter case, the thigh will be forcibly flexed by the Y, and either adducted or abducted, while the head descends even to the tuberosity or perinæum, afterwards, perhaps, returning to be lodged near the thyroid foramen or on the dorsum. But if the Y be wholly broken, the limb is suspended by the remaining and comparatively slender capsule, which in such a case would probably be ruptured, thus abandoning the limb to the muscles. Of these, the psoas and iliacus offer a resistance most resembling that of the capsule, and produce an imperfect flexion. The biceps and other extensors may in certain positions interfere with flexion, as they doubtless do in the regular dislocation downward, while the adductors and flexors are also put upon the stretch when the limb is extended or abducted. It has been elsewhere stated that the muscles inserted immediately about the hip are subjected to the very powerful leverage of the femur, and are readily ruptured when unsupported by the ligament of the capsule. The same is true in a less degree of the long muscles, which are liable to laceration from the great violence necessary to sever the entire capsule. When this happens, the bone may be considered as fairly torn from the socket,—a grave accident, which rarely occurs, and in which the limb assumes no uniform position. The head of the bone might possibly, in such a case, be found on the tuberosity or in the perinæum, even when the limb is extended.

cause the operator was able to "elongate or pull down the limb" after reduction: a possibility supposed to depend on a fracture of the socket, but which may have resulted, if correctly reported, from a rupture of the Y ligament.

REDUCTION.

An irregular dislocation, with rupture of the Y ligament, if not the whole of the capsule, cannot be reduced by any rotation which depends for its efficiency upon the integrity of these ligaments. On the other hand, their ligamentous fibres can no longer interfere with a direct traction of the femur towards the socket, aided by local guidance, if required.

SPECIAL CONDITIONS OF DISLOCATION.

OLD DISLOCATIONS AND THEIR REDUCTION.

COOPER[1] cites a case of dorsal dislocation said to have been reduced after the lapse of five years by a fall from a berth on shipboard. Such an occurrence is by no means impossible, but would depend upon the condition of the acetabulum, and of the head of the bone, the changes in which would be influenced by the age and tendencies of the patient.[2] So long as the socket was still excavated, and the bones were not deformed by osseous growths, I should feel quite confident of breaking any adhesions, lacerating the newly formed capsule, and replacing the bone, by the great power of the femoral shaft as a lever, and of the flexed leg in rotating the head of the bone around the main ligament.[3] I am

[1] Op. cit., Case LXIV.
[2] For a case of dorsal dislocation reduced after eight months, see p. 55.
[3] Since the above was in type I have met with the following passage corroborative of the views here advanced, although its writer does not recognize the capsule as a source of resistance to reduction.

"It is doubtful if the capsule is ever an obstacle to the return of the dislocated bone. Certainly the altered shape of the head of the

unable to understand why Malgaigne, as quoted by my distinguished friend, M. Broca,[1] in the discussion elsewhere alluded to, should assign an indefinite period of two years or more as the limit for reducing a dorsal dislocation, and only fifteen days for that upon the ischiatic notch. By the flexion method, the latter luxation should, theoretically, be reduced with even more facility than the former, and after as long an interval.

A difficulty that may be seriously considered is the risk of breaking the femoral neck, if it has undergone fatty degeneration or atrophy from long disuse; and it might be well, in such a case, to rely rather on traction or other force exerted longitudinally upon the bone, than on rotation, where, from the immense power thus laterally applied, the neck is taken at great disadvantage. The angular traction, to be hereafter described, would be especially suitable, — although, from the greater facility of such an application of power, a better result might be anticipated in a dorsal or downward than in an old pubic or even thyroid luxation. Yet should fracture of the neck, or separation of a previous fracture, occur during such attempts at reduction, it may be fairly said that the patient will generally have a better limb, after its

bone never can prevent the return of the head to its articular cavity. And it is probable, that, where the articular cavity is partially obliterated, it is the result of extraordinary violence and consequent inflammation. I have found the cotyloid cavity retaining its depth and covered with cartilage after the head of the femur had been dislocated for three years. And Fournier has placed a dissection on record where the head of the femur had been dislocated during thirteen years, and in which the acetabulum retained its form and depth and cartilage. (Bulletins de la Société Anatomique, 1855.)" — On the Reduction of Old Dislocations, by Bernard E. Brodhurst, Assistant Surgeon to the Hospital : St. George's Hospital Reports, Vol. III., 1868, London.

[1] Union Méd., No. 79, p. 57.

inversion has been thus corrected, than with an unreduced luxation.

An illustration of these points is afforded by a case in the Chelsea Marine Hospital, under the charge of Dr. Graves. The patient, a man twenty-three years of age, about six months before entering the Hospital, had fallen from the mast-head, seventy-five feet, striking the thwart of a boat, which was broken by the fall, and dislocating his left hip. No attempt was made, at the time of the accident, Nov. 8, 1862, to reduce the displacement. At the time of entering the Hospital, the patient was wholly unable to walk, being carried, and placed in bed, where he remained. The limb was shortened about two inches, slightly flexed, and inverted to such a degree that the patella faced the inside of the opposite thigh, and the toes of the affected limb were more easily placed behind the heel of the other foot than upon the instep. The patient could partially flex the thigh, and also extend it nearly flat upon the bed, and could rotate the limb inward, but could not evert it. The head of the bone was readily felt upon the dorsum. Dr. Graves having kindly placed the man under my charge for the reduction of the dislocation, I flexed the limb once slowly upward upon the abdomen, — a movement which was attended with a continued fine crepitation about the hip. Upon examination, the head of the bone was now felt to be detached from the neck, and freely movable, like a grape-shot, among the muscles of the haunch. The patient was thereupon placed in bed, the position of the extended limb being much the same as before manipulation. In the course of a week the foot was gradually everted, after which extension was applied, and maintained during three months, being increased by degrees from seven pounds to about twenty-one. In two weeks from this time the patient began to move about on crutches, which after six weeks more were abandoned, and at the end of two years he was able to walk without a cane. The limb is now, six years after the accident, an inch and a half shorter than its fellow, but otherwise in proper position, and moves freely in all directions, although it cannot be everted much beyond the perpendicular. The head is firmly attached to the femur behind the trochanter, and seems with the latter to cover the acetabulum. The manipulation, in this case, was conducted in the presence of a considerable number of medical gentlemen, and the manner in which the head was detached from the shaft left no doubt upon their minds that the neck, as the result either of an original fracture or of subsequent inflammatory action, had not its normal strength. On the other hand, the present condition of the patient is much better than it would have been, had not the dislocation been treated. He walks freely and firmly, with but little lameness, runs up and down stairs, and can swing the limb in all directions.

Dislocation from Hip Disease.

In the dorsal luxation which follows aggravated hip disease, the anterior part of the capsular ligament usually supports and inverts the shortened limb: on the other hand, the head of the femur, which rests upon the dorsum of the ilium, produces, when disintegrated by disease, less inversion than if it were of normal size. Again, the displacement is generally a sub-luxation: but it may sometimes be complete. In a case of hip disease, occurring in a boy about ten years of age, which terminated fatally, I excised the head of a femur (the first instance of this operation in the United States) that was completely dislocated upon the dorsum.

The following is an instructive case of dislocation, perhaps connected with hip disease, and reduced by manipulation.

The patient was a feeble and slender boy thirteen years of age, who was said to have dislocated his hip upon the dorsum by a fall upon a barn floor about three months before, and whom I was requested to see in consultation. The head of the bone could be plainly felt upon the dorsum, the limb being as usual inverted, shortened, and a little flexed. I found, that, in abducting the limb after it was flexed, a very considerable force was required to raise the head over the socket, and still more in outward rotation to make it enter, which it did only after the capsule and other attachments had been freely lacerated. After reduction, the head of the bone readily and repeatedly escaped, and could be kept in place only by the expedient, elsewhere alluded to, of confining the limb. The foot was secured to the inside of the sound knee, and the limb, thus flexed, was abducted down to the level of the bed, where it was bound to the side of the bedstead by a folded sheet under the knee. In this constrained position of flexion, abduction, and eversion the patient remained for two and a half weeks, when I again saw him, and found the bone in place. But soon the hip-joint became stiff and painful, and sinuses slowly formed and opened in the groin, as if from hip disease. Upon inquiry, it was ascertained that the child had suffered from pain near the hip after a fall the preceding year, and had also lately recovered from protracted and grave disease of the bone near the ankle. The dislocation may or may not have been facilitated by this tendency to disease of the bone, but there can be little doubt that serious inflammatory action was awakened by the presence of the reduced femur in the socket.

Dislocation of the Hip, with Fracture of the Shaft of the Femur.

Cases have been reported of fracture, even of the upper third of the shaft, in which an accompanying dislocation was reduced by manipulation. There seems to be no good reason, why, after the firm application of lateral splints to the thigh, the attempt should not be made with entire success,[1] reliance being especially placed upon flexion and the local management of the head of the bone, which may be guided into its socket by the hands of the operator applied directly to it, or by a towel in the groin. Angular extension of the lower fragment of the femur may draw upon its upper muscular insertions, and likewise make room for the upper fragment to follow it; but it is obvious that nothing can be effected by its rotation.

Spontaneous Dislocation.

Cases have been cited of individuals who could partially luxate and reduce the head of the thigh-bone at will, by the action of the muscles of the hip. Hamilton has collected three such cases.[2] I have had an opportunity of examining two, and Dr. Lyman, of Boston, has communicated to me the details of a third, all of which were dorsal luxations.

In the first of these cases, — that of a soldier under the charge of Dr. Langmaid, to whom I am indebted for the opportunity of examining it, — the hip was dislocated while the legs were crossed, a wagon in which the man was riding having pitched into a hole. In a few hours the hip was reduced by flexion.

---

[1] See Hamilton, Op. cit., p. 666.   [2] Op. cit., p. 644.

Eight days after the accident, in attempting to walk upon the limb, it was again partially luxated, when the patient himself replaced it by pushing against it with one hand and pressing with the other against his knee. Since that time both luxation and reduction have been comparatively easy, and the patient now displaces the head of the bone backward upon the edge of the socket by muscular action, and reduces it by "throwing the leg out sideways." The luxation is sometimes attended with pain, and the prominence caused by the head of the luxated bone is sensitive to the touch. In this and the following case, the displacement is rather a sub-luxation, and the limb exhibits slight flexion, shortening, and inversion.

In the second case, — that of a gentleman formerly of Boston, — the phenomena are much like those just described: the bone being slipped out and in upon the dorsal edge of the socket by muscular action at will.

A third case was under the care of Dr. E. M. Moore, of St. Mary's Hospital, Rochester, N. Y., who has published photographs of it, from which the annexed figures are taken. The following account of this case has been kindly furnished me by Dr. G. H. Lyman, of Boston, who obtained it from Dr. Moore.

"John B. Parker, private, Co. H, 148th New York Volunteers, while on the march from Bermuda Hundred to Drury's Bluff, May 13, 1864, was skirmishing up a hill, and sprang back suddenly to avoid the gun of a comrade in advance. His left foot became entangled, and his weight dislocated his hip. He felt the injury, and supposed it out of joint. Some comrades pulled it in. He immediately resumed his skirmishing, and marched seven miles, from 10 A. M. till 6 P. M. He lay down at night, and went on duty the next day, sharp-shooting, crawling all day. He continued this kind of duty five days, and returned to camp, when he was immediately put on intrenchments, and worked two days and nights. Afterwards he went on picket, and entered the hospital May 28. At present he can luxate the hip-joint at any time, and does it by pressing the foot on the floor to fix it firmly, contracting the adductors, and throwing out the pelvis. The head suddenly leaves the acetabulum, and goes on the dorsum ilii."

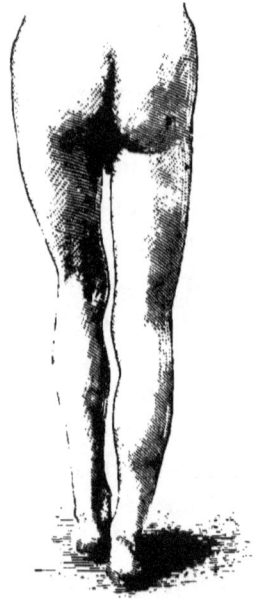 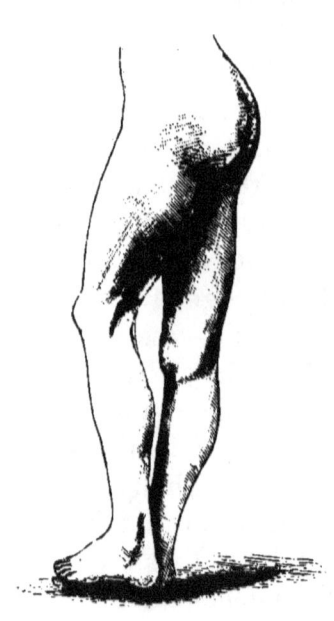

Fig. 41.   Fig. 42.

Although the lateral displacement and slight inversion show that this is only a sub-luxation, with the head upon the edge of the socket, yet the flexion of the limb, due to the elasticity and comparative integrity of the living tissues, makes it perhaps a better representa-

---

Figs. 41, 42. — Spontaneous luxation of the thigh. Dr. Moore's case.

tion of a common dorsal luxation than Fig. 4, which was photographed from the dead subject, and where the limb was purposely extended as far as the Y ligament would allow.

ANGULAR EXTENSION.

POUTEAU[1] first remarked upon the disadvantage of traction with counter-extension in the perinæum, which brings the thigh into a straight line with the trunk. Most surgeons have observed the tendency of the pelvis, when pulleys are used, to escape from the counter-extending bands in the direction of the applied traction. It is believed that the apparatus here described will be found efficient, both in confining the pelvis and in enabling the operator to apply extension to a limb which has been flexed for the purpose of relaxing the Y ligament. Lateral extension, with or without pulleys, can then be made in any desired direction by a towel passed round the head of the femur.

The patient being laid upon his back, the pelvis is secured to the floor by a T band passing across it laterally in front, between the superior and inferior spinous processes of each side, and vertically over the pubes and perinæum. The three extremities, each terminating in a strap and buckle, are fastened to the floor beneath the margin of the pelvis by common dislocation-hooks. The entire band, with the exception of its extremities, is cylindrical, about two inches in diameter, well padded and covered with buckskin. It firmly holds the pelvis by its pressure between the spinous

[1] Malgaigne, Op. cit., p. 867.

116 ANGULAR EXTENSION.

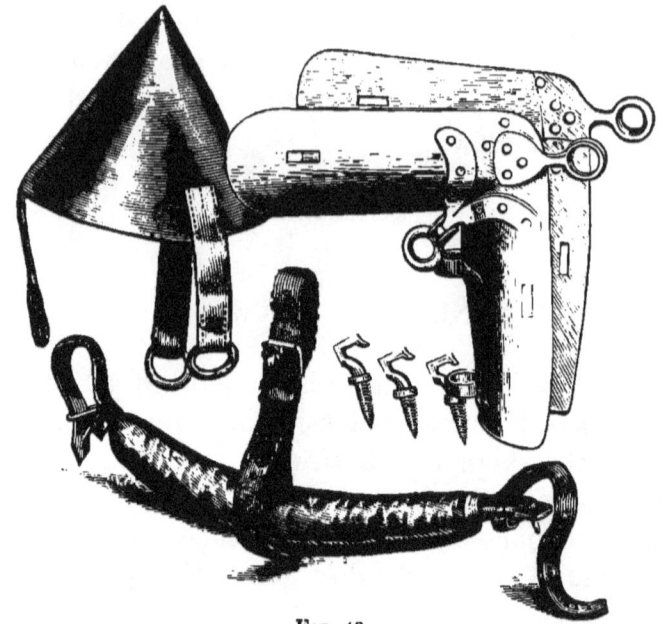

Fig. 43.

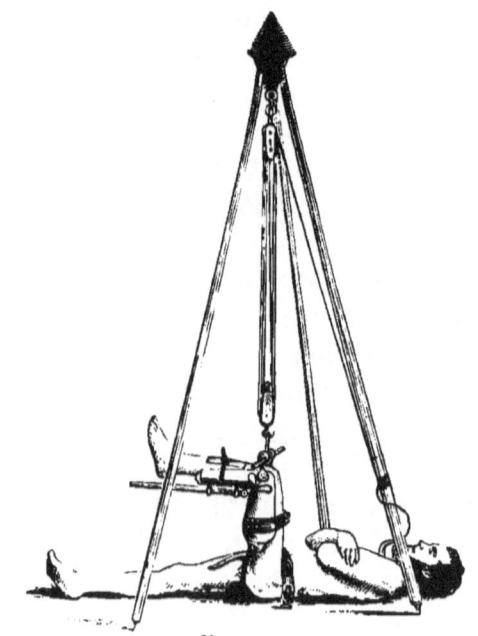

Fig. 44.

processes on each side and upon the pubes. To apply it, the three pointed hooks are screwed into the floor, one near each trochanter, and one near the perinæum; the band is then adjusted, and the pelvis buckled to the floor, after which it will be found that the thighs can be freely flexed. A tripod is now erected over the pelvis, consisting of three stiff poles about eight feet high, and held together at the top by a conical leather cap, with three short dependent straps and rings from which the pulleys are suspended. It remains only to attach the pulleys to the limb. This is effected by means of a strong right-angled splint of sheet-iron, extending nearly from the hip to the ankle, made concave so as to embrace the under surface of the thigh and leg, and padded, within which the limb, flexed at right angles, is confined by bandages or straps. Two iron rings riveted to the splint near the condyles of the femur receive a wooden rod about two feet in length and an inch in diameter, which crosses the ligament of the patella transversely above the head of the tibia; and to this rod, between the rings, the pulleys are attached by a strap or cord. Vertical traction is thus made exactly in the axis of the shaft of the femur.

Powerful rotation can be made by grasping the extremities of this transverse rod, while another useful movement, called by the French *bascule*, or tilt, may be effected by a similar rod in the axis of the leg below

---

FIG. 43.—Apparatus for angular extension. This wood-cut represents the conical leather cap and rings,—the angular splint, with rings above and below the knee for the passage of a transverse wooden lever, and of a longitudinal one beneath the calf,—the padded **T** pelvic band, and the hooks to attach it to the floor.

FIG. 44.—Angular extension. The pelvis is buckled to the floor. The flexed leg is suspended from the cap at the summit of the tripod by pulleys which are attached to a transverse wooden rod across the patella. This rod passes through rings on the angular splint, and serves to rotate the limb. A similar rod is seen beneath the leg.

the knee, passed through two rings beneath the splint, one near the ham, the other near the heel, beyond which it projects a foot or more, to afford a handle. By vertically raising this rod at its extremity we carry the head of the bone from the dorsum or pubes in the direction of the tuberosity.

Oblique extension may be made by changing the position of the tripod.

Although the need of this apparatus may be rare, it will prove occasionally efficient in reducing a luxation of long standing or complicated with fracture. At any rate, I cannot believe that the period is remote when longitudinal extension by pulleys to reduce a recent hip luxation will be unheard of.

FRACTURE OF THE NECK OF THE FEMUR.

IMPACTED FRACTURE OF THE BASE.

THE injury known as the "impacted fracture of the neck of the thigh-bone" has been well described by various writers. When it occurs, the neck, broken at or near its broad insertion into the head of the shaft, is driven into the loose cancellated tissue of the latter, and so fixed there that it sometimes requires a considerable force to withdraw it. That it may be a severe lesion, especially in the latter part of life, the numerous recent specimens to be found in museums sufficiently attest. In my own observation, while it is at least as frequent among elderly people as fracture of the neck within the capsule without impaction, the accident is comparatively common in middle life, and even later, and the bone is sometimes capable of uniting after a few months, with little deformity.

This fracture is characterized by shortening and eversion of the limb, sometimes so inconsiderable that we are obliged to accept a diagnosis based upon an almost imperceptible eversion, and a shortening of half an inch or less, by careful measurement.

The Museum of the Massachusetts Medical College contains a valuable collection of impacted fractures of the hip; and having, through these specimens, become familiar with the eversion exhibited by them in various degrees, I had my attention more carefully directed to the subject by the following not unusual case. A gentleman slipped upon the ice before his door, and fell upon his hip. He walked up stairs with assistance, and was placed upon his bed. His attending physician, in the absence of any obvious shortening or eversion of the limb, entertained some doubt in regard to the nature of the injury, but, after ten days, finding no improvement in the symptoms, — the pain and soreness having in fact increased, — requested me to see him. The local tenderness and pain on motion, together with a very slight eversion — best seen on attempting to invert comparatively the two feet — and a shortening of less than half an inch, led me to the conviction that the bone was slightly impacted; and I conceive this view to have been corroborated by callus subsequently felt about the trochanter, and by the length of time required for the recovery, — the patient having been confined to his bed a little more than two months, and unable to walk without crutches until after the lapse of four months.

Since that time I have had sufficient opportunities to satisfy myself, that, though this accident may be serious when it occurs late in life, it is by no means so to a middle-aged and healthy subject,—that the impaction is sometimes slight, and its indications proportionably so,—and that the following signs may be relied on as generally pathognomonic: disability; pain and tenderness resulting from local violence, especially when applied laterally, as in a fall upon the hip; shortening and eversion, however slight; absence of crepitus; and lastly, the rotation of the trochanter through an arc of a circle of which the head of the bone is the centre, instead of upon the axis of the shaft, as in detached fracture of the neck.

The practical importance of readily identifying this fracture lies in the fact that its progress, as regards both time and good union, is in general more favorable than that of the unimpacted fractures,—that, though it is a comparatively common and disabling accident, it may exhibit little deformity,—and lastly, that the object of extension in its treatment is to steady the limb, and not to draw it down.

The following details of the anatomical structure of the femur sustain the foregoing statements in respect to the shortening and eversion incident to this lesion.

ANATOMICAL STRUCTURE OF THE NECK OF THE FEMUR.

Let a well-developed femur be placed in a vice with its back towards the observer, in its natural upright position, but obliquely, as if the legs were widely separated, the shaft being so far inclined that the neck is horizontal. Let a first slice be now removed from the top of the head, neck, and trochanter by a saw carried horizontally through the neck. Let a second and third slice be removed in the same way, so that the neck shall be divided into four horizontal slices of equal thickness.[1]

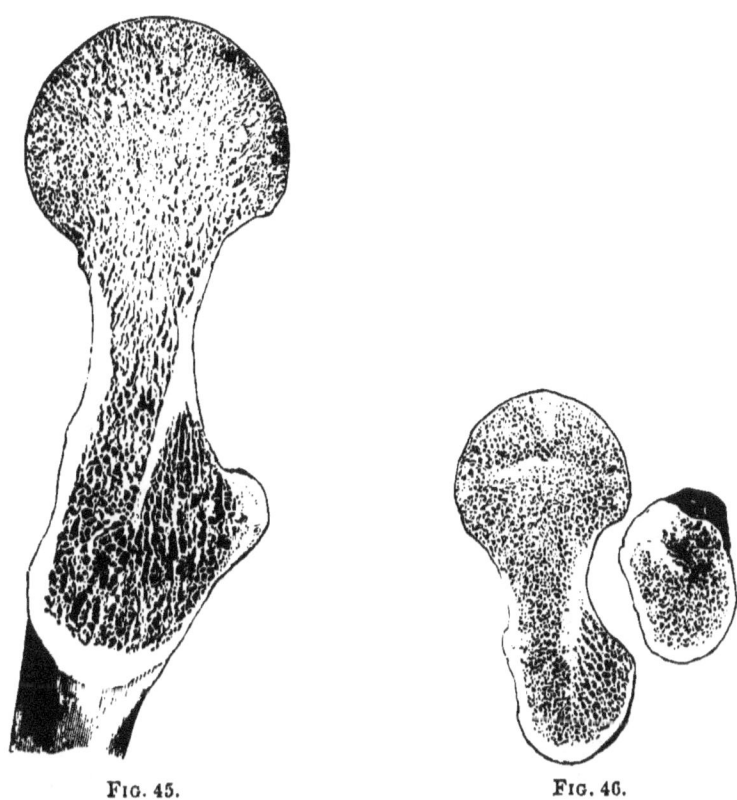

Fig. 45.   Fig. 46.

[1] If the head of the bone be now vertically transfixed by a wire, the sections may be spread for examination, like a fan.

Fig. 45. — Exhibits a bird's-eye view of a horizontal section of the neck of the femur, showing the posterior wall plunging beneath the intertrochanteric ridge, at the angle where the neck joins the shaft. The posterior wall is of the thinness of paper, and here impaction occurs. The anterior wall, on the contrary, is seen to be quite thick, and forms by its fracture a hinge which is very rarely impacted.

Fig. 46. — The same. The section of the shaft near the lesser trochanter shows the lower extremity of the septum, where the wall is thicker and changes its direction. (From a photograph taken in 1861.)

It will be found that the upper section exhibits the anterior and posterior walls of nearly equal thickness, —but that, as we approach the lower surface of the neck, the anterior wall becomes of great thickness and strength, while the posterior wall remains thin, especially at its insertion beneath the posterior intertrochanteric ridge, where it is of the thinness of paper.

Rotation.

The result of this conformation is obvious. In impacted fracture the thin posterior wall is alone impacted, while the thick anterior wall, refusing to be driven in, yields only as a hinge upon which the shaft rotates to allow the posterior impaction. This phenomenon, varying a little with the injury, is constant in every specimen of simple impacted fracture I have examined; and in fact it must be so, from the arrangement of the bony tissues, which at once invites and explains the eversion.[1]

[1] M. Robert, in a memoir upon impacted fractures of the neck of the femur, attributes the posterior impaction to the supposed fact that the tangential plane of the external surface of the trochanter is inclined obliquely backward to the axis of the neck, and that a force applied to its centre would tend to increase the obliquity of this angle, and thus to produce outward rotation of the shaft. The shortening of the limb he attributes to the fact that the impaction is greatest at its lowest part. (Mémoire sur les Fractures du Col du Fémur accompagnées de Pénétration dans le Tissu spongieux du Trochanter, par Alph. Robert, Professeur Agrégé, etc. : Mém. de l'Académie de Médecine, Tom. XIII. p. 487.)

Shortening.

The hinge before alluded to is oblique, following the anterior intertrochanteric line. Were it vertical, by bending this hinge we should produce rotation without shortening. On the other hand, if it were horizontal and transverse, bending it would produce shortening without rotation. But as it stands at an angle of 45°, the shaft rotating upon this broken interval is shortened in proportion to its rotation, — or, what is the same thing, the neck is reflected upon its hinge downward and backward, till its axis, normally oblique, may become even transverse, with great outward rotation of the shaft and a shortening of perhaps two inches. This is probably the most common cause of shortening, although the head of the bone may be otherwise depressed.[1]

True Neck.

Upon examining the lower of the above sections in a well-marked bone, the posterior or papery wall of the neck will be seen to be prolonged by radiating plates into the cancellous structure beneath the intertrochanteric ridge. That the thickest of these (Fig. 45) is a continuation of the true neck may be shown in another way. Let the whole of the posterior intertrochanteric ridge, including the back part of both trochanters, be removed by a narrow, thin saw. (Fig. 47.) The bone being now laid upon a table, let a chisel, or, what is better, a gouge, be held perpendicularly upon the cancellous structure thus exposed, and lightly twirled until the friable and spongy tissue is removed and the

---

[1] See pp. 129, 134, 136.

instrument arrested by the septum or wall alluded to. To expose its inner surface, the shaft should be split by a vertical and curved section behind this wall, and the cancellous structure removed in the same way.

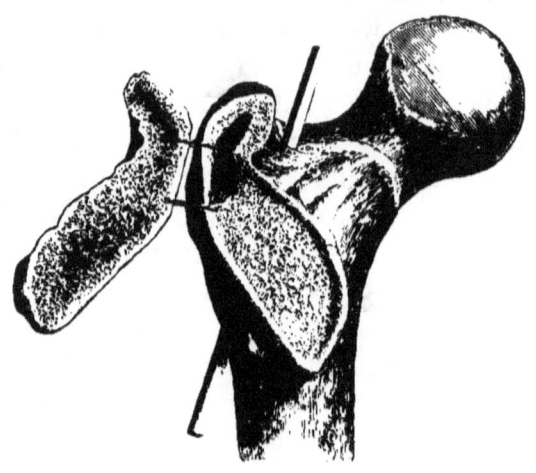

Fig. 47.

The septum will then be distinctly seen, as a thin, dense plate of bone continuous with the back of the neck, and reinforcing it, plunging beneath the intertrochanteric ridge in an endeavor to reach the opposite and outer side of the shaft. At its lower extremity it curves a little forward, so as to take its origin, when on a level with the lesser trochanter, from the centre, instead of the back, of the cylindrical cavity, — a disposition easily seen in a transverse section of the shaft just above the trochanter minor. (Fig. 46.) Or it may be said that the posterior wall of the neck forks before reaching the intertrochanteric line, one layer

---

Fig. 47. — Anatomy of impacted fracture. The intertrochanteric ridge has been removed, and the cancellous structure so excavated as to exhibit the true neck beneath. The rod is placed in a longitudinal fissure by which the shaft of the bone has been split, in order to exhibit the true neck from within. (From a photograph taken in 1861.)

being seen upon the surface, while the other dives beneath the intertrochanteric ridge in a vain attempt to reach the outer wall of the shaft. If these views be correct, the intertrochanteric ridge is simply a buttress erected for the insertion of muscles upon and over the true neck, by the impaction of which it is in fact often split off and detached in a mass, the force exerted by the true neck, though slight, being nevertheless an effort to resist such impaction.

REMARKS.

Surgical writers have been at some pains to indicate the distinguishing marks and tendencies of the so-called fractures "within" and "without the capsular ligament," — names which have but little practical significance. While the impacted fracture of the base of the femoral neck unites by bone, if at all, there seems to be a decreasing tendency to osseous union as we approach the smaller portion of the neck near its head, — a circumstance probably due in part to the feeble nutrition of the detached extremity, and in part to its mobility. The fact, which Sir Astley Cooper did not deny, that bony union is possible "within the capsular ligament"[1] and at the slenderer portions of the neck, is now sufficiently attested by existing specimens, our College Museum possessing two of these. But in examining specimens of such bony union, it is often difficult to say just how far the fracture was originally within or without the capsule, because the exact position and limit of the capsule itself are variable;[2]

[1] Op. cit., pp. 137, 138.
[2] The Insertion of the Capsular Ligament of the Hip-Joint, and its Relation to Intra-Capsular Fracture, by George K. Smith, M. D., Demonstrator of Anatomy, etc., New York, 1862.

and if we except the impacted fracture at the base, it is impossible during life, by any justifiable examination, to decide what part of the neck is broken, or whether the fracture has occurred within or without the capsule. Nor is it a matter of importance in the treatment, which is one and the same in both cases,— or in prognosis, if the so-called varieties cannot be distinguished. In lecturing upon this subject, I have been in the habit of dividing the injuries of the neck of the femur into the impacted fracture of the base of the neck and the unimpacted fracture of the rest of the neck, without regard to the capsule,— a practical classification, embracing a majority of cases, and to which the other lesions may be regarded as exceptional.

It is, indeed, possible for the small extremity of the neck to be impacted into the detached head, and so steadied by it as to favor union. Such was the injury in specimens in our Museum, described below. It is also possible for the base of the neck to be impacted with inversion. But in the large majority of cases, if there is a serious injury to the neck of the bone, it is either a common impacted fracture of the base of the neck, easily diagnosticated by the signs already described, or some other fracture, about which it is of no practical consequence in its treatment to know anything, except that it exists and needs extension.

In brief, the presence of excessive pain on motion leads to the suspicion of severe injury. The age of the patient,— the shortened, everted, loosely hanging, and uncontrollable limb,— crepitus, which, when once felt, is as satisfactory as if felt, to the detriment of the patient, many times,— and lastly, the head of the trochanter rotating on the axis of the shaft, and not through an arc, readily and quickly identify the unimpacted fracture. On the other hand, the impacted fracture of the base, which occurs in the adult at all ages, though more frequently in the latter half of life, is characterized by less local pain and disability, by shortening and eversion[1] which may be slight, and by the absence of crepitus, while the trochanter rotates through an arc upon the articulation as a centre.

The importance of distinguishing between the different fractures of the neck of the femur is not so great as to justify any protracted or considerable examination. Flexion of the thigh, its repeated rotation, or other unscrupulous or unskilful handling, is liable to lacerate the remaining capsule, to displace the bony fragments, or, by loosening and detaching an impacted fracture, to render its union more difficult, — adding, perhaps, to the accuracy of the diagnosis, but directly diminishing the chances of the patient.

The treatment of all these fractures is similar, the unimpacted fracture obviously requiring extension, the purpose of which, in the impacted fracture, is to steady, not to elongate, the limb. Among the many expedients presented to the choice of the surgeon, I have for my own part found as good results, even in bad cases, from a flat bed, with a book or other weight attached to the foot for extension, and perhaps a broad band about the hips to steady the parts, and a cushion or pillow under the broken hip to prevent its eversion, as from more complicated and less comfortable apparatus. The prognosis of these fractures it is difficult to give. Elderly people may die of them at the end of a few weeks, or may linger many months. On the other hand, when the fracture is near the base of the

[1] As before stated, a slight eversion is perhaps best indicated by a comparison of the extent to which the two limbs can be inverted.

neck, cases occur of recovery, with little lameness, both from the impacted and the unimpacted varieties, especially the former.

To facilitate a differential diagnosis, the principal lesions of this region will now be described.

Impacted Fracture of the Base of the Neck with Inversion.

This accident is of rare occurrence. Smith and Hamilton each cite but one case. Indeed, the structure of the bone, as has been shown, is such as to insure an almost uniform eversion of the shaft. A specimen from a dissecting-room has enabled me to examine this rare lesion, and to identify the conditions under which it probably occurred. In this subject, an old woman, the limb was flexed a little, shortened to the extent of three inches, and inverted so that the patella faced inward; the limb was in slight abduction, and could neither be everted nor brought to the median line. The trochanter was felt to be much thickened. Upon examination of this exceptional specimen, the neck of the bone was found to be firmly united at right angles with the shaft, which was split open and spread so widely as to receive the whole

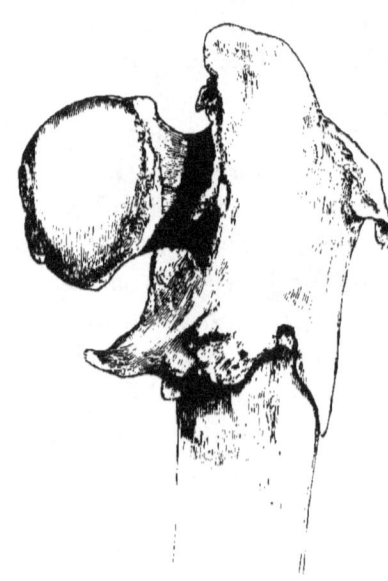

Fig. 48.

impacted neck, leaving a fissure an inch or more long and a quarter of an inch wide between the anterior wall and the neck, and extending nearly to the outer wall of the shaft, while another similar fissure exists behind the neck. The principal posterior fragment comprised the two trochanters with the intertrochanteric ridge, and also a large fragment of the external portion of the shaft, — while above, the region of the great trochanter seemed to have been comminuted and driven downward and inward. Anteriorly, the fracture had occurred, as usual in impacted fracture, along the oblique spiral line, although differing from that injury, the neck being deeply driven in behind this wall, from which it had slipped, instead of turning upon it as a hinge. The whole upper part of the bone above the trochanter minor seemed to have been bent inward, so that the posterior intertrochanteric line, instead of inclining obliquely to the axis of the shaft, was vertical. The inversion was due to the extent of the comminution, which had separated the walls of the shaft so as to receive in the interval the whole neck, instead of the posterior wall only, as commonly occurs, thus producing an anterior as well as a posterior impaction. The shortening resulted both from the horizontal

Fig. 49.

Figs. 48, 49. — Impacted fracture with inversion. Fig. 48, posterior view; Fig. 49, anterior view. The foreshortening fails to show the length of the tapering prolongation of the trochanter minor.

position of the neck, and from an additional upward displacement of the shaft, caused by the comminution. A good deal of callus had been thrown out in various directions, and the movements of the limb must have been quite restricted. A curious spicula stood at right angles with the shaft near the lesser trochanter, and may have been a displaced fragment, or the ossified insertion of the psoas tendon. The same bony spicula exists in another specimen before me, and is not uncommon.

In examining the accompanying illustrations, it will be seen that the intertrochanteric ridge is split off, as often happens, but in this case it has carried with it the outer and posterior walls of the shaft, with the two trochanters.

Smith[1] (Case XLVI.) cites a similar case of inversion, which the accompanying figure shows to have resulted from a similar cause. The posterior intertrochanteric ridge, with the greater part of the two trochanters, has been detached in a mass, and so widely, that the neck of the bone has slipped from its anterior hinge. In both specimens the impaction is arrested near the outer wall of the shaft. The entire neck in my specimen can be seen through the lateral fissures, while in that of Smith its extremity is detected through an interval of the fragments near the great trochanter. A similar specimen, numbered 248, in the Mütter Museum in Philadelphia, shows neither inversion nor eversion.

---

[1] A Treatise on Fractures in the Vicinity of Joints, etc., by Robert William Smith, M. D., M. R. I. A., etc., Philadelphia and Dublin, 1850. (See also Case XXXVII.)

IMPACTED FRACTURE OF THE NECK OF THE FEMUR
NEAR THE HEAD.

The following cases of impacted fracture of the femur near the head, one resulting fatally, the other in complete recovery, with the exception of persistent pain, may be regarded as instances of fracture fairly within the capsule. They illustrate not only the possibility of bony union of the detached articular extremity, but also the circumstances which contribute most frequently to its occurrence, — if, indeed, they are not essential to it. In two specimens (Nos. 2111 and 1540), in the Museum of the Massachusetts Medical College, of undoubted bony union after fracture of the femoral neck, the line of separation is near the head, which is tilted obliquely downward towards the lesser trochanter, as in the following cases.

Case 1.

A man aged seventy-six entered the Massachusetts General Hospital, March 9, 1863, under the charge of Dr. Gay, who has kindly furnished me with a record of this interesting case. The patient fell in the evening upon the sidewalk, striking the right trochanter. Feeling only that he had received a severe bruise, he crawled up stairs alone, and sat in his chair long enough to read his newspaper before going to bed. Two days after, he entered the Hospital. Upon examination, it appeared that the right leg was shortened half an inch; the foot was everted, and could not be inverted beyond the perpendicular; the thigh could be flexed and extended without difficulty, but with pain; the trochanter was less prominent than that of the other side. At the end of two weeks he died of pneumonia, at two o'clock in the afternoon; but at half past ten in the morning of the same day he had asked to have the splints removed, saying that the leg felt well, at the same time lifting the whole limb several inches from the bed without assistance.

In this interesting case, of which an excellent illustration is here given, the head was found to be broken from the articular extremity of the neck, which was short and thick, the fracture behind being almost at the line of junction of the articular cartilage and the bone, while in front it ran irregularly across the neck from a quarter to half an inch below this line. The head was bent on the neck obliquely backward and downward towards the lesser trochanter,—the tilting of the head opening the fracture on the outside of the neck,—and was so firmly impacted that considerable force was required to withdraw it. The impaction was double, the shell of the neck being driven to the

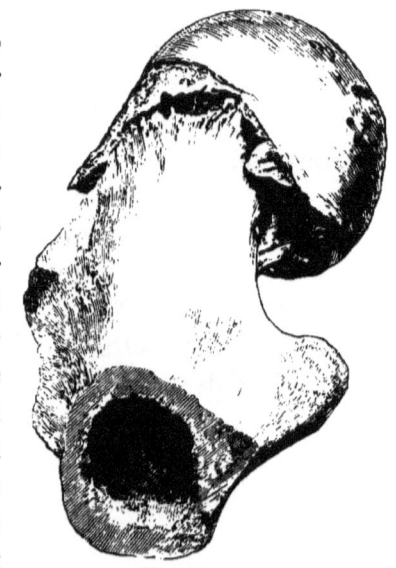

Fig. 50.

Fig. 50.—Dr. Gay's case of impacted fracture near the head. In this specimen the neck of the bone was originally short and stout. Below and behind it the fracture follows, as nearly as may be, the line of the articular cartilage, while anteriorly and above it is about half an inch distant from this line. The impaction in the recent state was firm, the thin surface of the neck at its lower and posterior part having been driven into the cancellous structure of the head to the depth of about half an inch, while the cancellous structure of the head of the bone has penetrated to the depth of three quarters of an inch into that of the neck, this mutual impaction being very firm. The head rests obliquely upon the lower fragment, as if the shaft had been rotated outward, opening the anterior part of the fracture to the width of nearly a quarter of an inch.

depth of half an inch into the head behind, while the centre of the head had entered the cancellous tissue of the shaft, being much the more dense of the two. The patient was evidently not aware of the existence of fracture, and it would have been impossible for the surgeon to infer before death the exact nature of the injury. It is fair to suppose that two bony fragments thus mutually impacted and held in apposition would have united by bony union, had the patient lived; and in this case it cannot be doubted that the fracture was wholly within the capsule. An additional interest attaches to this specimen in connection with the subjoined case of fracture, almost identical with it in character, and presenting unequivocal bony union.[1]

Case 2.

The following case of bony union, in a fracture curiously resembling the preceding, occurred in the practice of Dr. Cushing, of Dorchester, Mass., a practitioner of large experience, whose opinion in respect to the general character of an injury of this sort would be entitled to weight, even were it not corroborated by the specimen here represented, the section of which shows unequivocal evidence of fracture. A woman seventy years of age, while reaching to wind up a clock, fell upon her side. Dr. Cushing, being called at once, found, that, although the limb was not obviously displaced, it was so disabled as to leave no doubt of the existence of a fracture. The patient was laid upon

[1] For a case of mutual impaction of the neck and head, but complicated with a second impaction, old or recent, of the base of the neck, see a paper by Thomas Bryant, F. R. C. S., etc., in the Medical Times and Gazette, May 1, 1869. As the result of this double impaction, there was "some shortening of the limb, but no eversion of the foot."

her back, with the knee flexed and two pillows beneath it. For two and a half or three months she kept her bed, and then began to sit up with the limb extended. Crutches were used for six months longer; then a crutch and a cane; but for the last two and a half years neither, the patient being able to go about the house and a little way out of doors. There was little, if any, shortening, and she limped but slightly. During the first few weeks she had much pain at the seat of the injury, and in the limb, which was gradually atrophied. Her health was generally good until near her death, four years and eleven months after the accident, from internal disease.

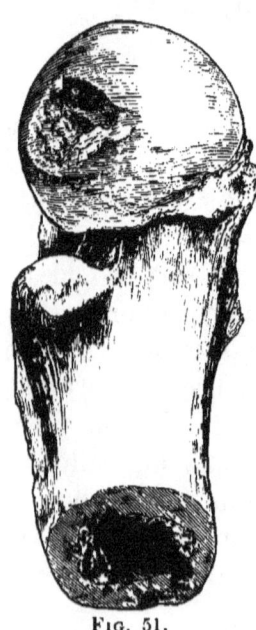

Fig. 51.

In this instance, also, the neck of the femur is short and thick,— the line of fracture corresponding very nearly with that of the articular cartilage. The head of the bone has been depressed so that the neck is now nearly transverse, — the head being also bent obliquely backward and downward towards the lesser trochanter, and the shaft thus rotated outward. In front the neck of the bone projects beyond the articular cartilage, while behind it is buried beneath it. As in the preceding specimen, the neck is thus posteriorly impacted into the head, which in bending backward opens a fissure in front, filled with an irregular bony callus.

It was not observed in this case that the limb was

Fig. 51.—Dr. Cushing's case of impacted fracture near the head, with bony union.

everted, as the specimen implies. At the time of death the foot was straight. Smith records a case (No. LVIII.) in which there was slight inversion.

Comminuted Fracture of the Trochanters without Impaction.

The alleged injury thus described detaches the entire neck from the shaft, and is generally represented in museums by specimens the comminuted fragments of which are reunited in their normal position. Museum specimens of this fracture are less frequent than those of the common impacted fracture, and might be still more rare, were it not that the impaction, which I conceive to be the rule in fracture at the base of the neck, does not always persist, being sometimes liberated by the extensive comminution of the bone, or by force subsequently applied. The impacted bones are undoubtedly separated, in many cases, from want of care both in the examination and in the subsequent treatment of the patient, and likewise in the process of their preparation as specimens.

It is fair to suppose that local crepitus can be felt in the trochanteric region, so extensively comminuted. The lines of fracture present great variety. The anterior and posterior trochanteric walls, or either of them, may be detached entire or in fragments. The posterior intertrochanteric ridge may be split off partially or wholly, and the trochanter minor broken off by itself. The summit of the outer trochanter, and in fact the whole upper region of the shaft, may be comminuted and driven in as by a blow from above.

As in other fractures of this part, inversion of the limb is here the rare exception, and eversion the rule.

Smith cites twenty-eight cases of extra-capsular fracture, of which four only were inverted. Assuming that anterior impaction is essential to inversion, we may seek the cause of the latter both in the direction of the blow received and in the action of the muscles. The influence of these is well illustrated in the case already detailed (Figs. 48, 49), where the mass of large and small rotators evert the upper fragment only, leaving the shaft to be inverted by the anterior fibres of the gluteus medius, and especially of the gluteus minimus, which is inserted lower down. This occurs when both the trochanters are detached, whether separately, or, as in this specimen, in a single piece. Smith's four cases of inversion[1] sufficiently illustrate these points, as does also a specimen in the Chatham Museum,[2] where, in addition, "an arch of new osseous matter" "extends from the anterior inferior spinous process of the haunch-bone across the joint to the upper part of the shaft of the thigh-bone," and which, it may be inferred from its insertions, was the Y ligament and the neighboring fibres. (See Fig. 22.) Shortening not unfrequently results from mere depression of the neck, without corresponding rotation of the shaft, as in the regular impacted fracture; and the transverse neck may then be displaced outward upon the shaft, so as to resemble a hammer upon its handle.

[1] Op. cit., Cases XXIX., XXXVII., XXXIX., XLVI.
[2] See the Third Fasciculus of Anat. Drawings, etc., Army Med. Museum of Chatham, — also, A Case of Fracture of the Neck and Trochanter of the Thigh-Bone with Inversion, etc., by Geo. Gulliver, Edinb. Med. and Surg. Journal, 1836, Vol. XLVI. p. 312.

Fracture of the Neck of the Femur resulting in False Joint.

The frequency of this lesion is attested by the common museum preparations, showing the hemispherical head of the bone slipping upon the absorbed and shortened neck, or upon a broader surface with more restricted motion nearer the shaft. In the latter case, the lower surface of the neck not unfrequently rests upon a bony projection near the lesser trochanter. False joint is a frequent result of unimpacted fracture, and is not to be averted by any special form of apparatus.

Crack in the Neck of the Femur.

It is obvious, that, while a simple crack or fissure of the femur would produce no immediate deformity, it might yet give rise, as in the radius at the wrist, to lameness and inflammation of long duration, with corresponding obscurity of diagnosis. The tendency of glass tubes and other brittle cylinders to crack in a spiral line is well known; and M. Gerdy has remarked upon the occurrence of oblique or spiral fissures in the long bones, producing at their intersection acute angles like the letter V. Those of the tibia sometimes exhibit a singular symmetry and mutual resemblance.[1] The annexed wood-cut (Fig. 52) represents a portion of a left femur from a patient who died under my care at the Massachusetts General Hospital, in one of my wards, of fracture and internal injuries. The specimen is now in the Massachusetts Medical

---

[1] See Pratique Journalière de la Chirurgie, par Adolphe-Richard, Chirurgien de l'Hôpital Beaujon, etc., Paris, 1868, p. 67.

College Museum, and has been described by Dr. J. B. S. Jackson, Dr. Mussey, and Dr. Hamilton. The femur is large and well marked. A spiral fracture ascends the shaft and winds round the neck, completely detaching it, except at a narrow isthmus in front, half an inch wide.

Fig. 52.

The shaft is broken transversely, eight inches below the trochanter. Here a spiral fissure begins, near the linea aspera, and winds upward and inward to the front of the bone, crossing the anterior intertrochanteric line midway between the trochanters; thence vertically upward to the outer edge of the cartilage; thence transversely across the top of the neck to its posterior surface, here touching the cartilage again; thence vertically down behind the neck to a point half an inch from the lesser trochanter, terminating on the under side of the neck in an S-shaped extremity, half an inch from the point where the fissure crosses the intertrochanteric line in front. The elastic bony pedicle thus formed allows a slight springing motion of the head, but maintains it firmly in place.

Fig. 52. — Crack of the femoral neck. Near the lesser trochanter is seen the hook-like extremity of the fissure, separated by a narrow interval of bone from the main line of fracture.

FRACTURE OF THE PELVIS.

The following remarks on fracture of the pelvis are introduced here, chiefly with the view of showing how far this injury may be mistaken for regular dislocation of the hip. With this view the subject has been divided into four heads, comprising, respectively. 1st, Fracture of the rim of the acetabulum; 2d, Fracture in which the head of the bone is driven through the acetabulum into the pelvis; 3d, Suspected fracture of the acetabulum; 4th, Fracture of other parts of the pelvis. A few cases are given in illustration of each of these lesions. The more instructive of these are, of course, such as have been verified by autopsy. But there are some which are authenticated only by well-marked crepitus, and perhaps mobility of the detached fragment; and it is then important that crepitus should not be confounded with the grating which results from the attrition of unbroken bone or cartilage. Finally, there are still others, and by far the most numerous, in which a fracture of the socket has been inferred only from a supposed impossibility of reducing the luxated femur, or of retaining it in place after reduction. It need not be said that these last cases are more conclusive to the observer than to the reader.

Fracture of the Rim of the Acetabulum.

To afford satisfactory evidence, cases of this sort should have been identified by autopsy, or at least by crepitus. Unfortunately, but a small part of the reported cases are thus elucidated, and fracture has been generally inferred, because the head of the bone could not be restored to the socket, or could not be kept there. It is probable, that, when the rim of the socket is broken on the side either of the dorsum or of the foramen ovale, the signs of the displacement do not vary materially from those of the regular luxations. The regular backward displacement, for example, may be complicated with a detached rim, which, if enough be left to engage the head of the bone, in no way interferes with its conditions as a luxation, except that the bone tends to slip backward after being reduced. The same principle probably holds true in the case of fracture of the rim on the side of the foramen ovale, and also of the upper part of the socket, unless the fracture involves the upper insertion of the Y ligament, in which case the detached fragment might be so displaced as materially to modify the position of the limb, especially so far as its flexion or inversion was concerned. Such a luxation would be irregular.

These displacements, especially the displacement backward, demand the usual attempts at reduction by flexion. Although the bone inclines to slip from the socket, it can be retained there, in cases of a sort heretofore considered difficult of treatment, by angular extension, with an angular splint attached to the ceiling, or some other point above the patient; or if any manœuvre has reduced the bone, the limb should be retained, if possible, in the attitude which completed the manœuvre.[1]

The following case occurred at the Massachusetts General Hospital, under the care of Dr. Gay. The patient, aged thirty-six, a robust and healthy man, fell from the roof of a building, striking upon the right hip. In the recumbent position the leg was shortened and inverted, the toes crossing the opposite instep. Being

[1] See p. 56.

etherized, the thigh could be flexed at a right angle with the abdomen, there being crepitus in the region of the neck of the femur. The limb, when drawn down, was still shortened half an inch. The patient having died of other injuries, the autopsy showed the head of the bone partially dislocated backward, and resting upon the posterior fractured edge of the socket, the whole posterior wall of the socket having been broken away in a mass. The detached fragment measured one and a half inches square. The posterior surface of the head of the bone was deeply indented by the fractured edge of the acetabulum, against which it had impinged after displacing the portion broken off. A transverse crack extended through the acetabulum from the upper sciatic notch to the foramen ovale. The position of the limb in this case did not differ from that in the usual partial dislocation behind the tendon, and was determined by the same mechanism.[1]

[1] In a case of dorsal luxation with inversion, reported by Maisonneuve (Clin. Chirurg., 1863, p. 168), the autopsy showed fracture of the posterior part of the border of the socket.

Sir Astley Cooper's Case No. LXXI. is one of regular dislocation below the tendon of the obturator internus, which tightly embraced the neck of the bone, with shortening and inversion of the limb, although the posterior part of the acetabulum was broken off, and there was other extensive fracture of the pelvis.

Dr. M. Tyer's third case was shown by the autopsy to be a regular backward dislocation with inversion, — the posterior and inferior margin of the acetabulum being detached, and displaced towards the coccyx.

On the other hand, in Dr. M. Tyer's first case, the limb was *everted* while flexed and shortened, an inch and a half of the rim being completely detached at the upper and posterior margin of the acetabulum. The remaining portion of the rim may not have been sufficient to turn the head backward, and thus compel inversion of the limb. In a second case, the toes crossed the tarsus of the other foot, and the autopsy showed a fracture of the upper margin of the rim of the

Fracture in which the Head of the Femur is Driven through the Acetabulum.

In regard to this accident, Hamilton well remarks: "There seems to be no certain rule in relation to the position of the limb; but it is found to take the one direction or the other, probably according to the direc-

acetabulum. (Glasgow Med. Jour., Feb., 1830; Amer. Jour. of Med. Science, 1831, Vol. VIII. p. 517.)

For a case of dorsal luxation with shortening, inversion, crepitus, and difficulty of retaining the reduced bone in the socket, see the work of Sir Astley Cooper, Case XXXIX.

In the following case of fractured acetabulum, the upper insertion of the Y ligament was detached. The patient, fifty-eight years of age, was caught by a revolving belt. The right limb was shortened a quarter of an inch, and so far everted and straight that the internal condyle of the left femur lay in the popliteal space of the injured one. The right groin was filled up. Towards its middle, and outside the femoral artery, was a hard, resisting, and obscurely spherical tumor, masked by the glands and swollen tissues. Flexion with outward rotation and local downward pressure failed to reduce the luxation; but on a third trial, flexion and downward pressure during slight abduction, instead of outward rotation, succeeded. Seven months afterwards, the death of the patient from another cause showed a united fracture of the socket, comprising the external and anterior third of the rim with the two anterior spinous processes of the ilium. (M. Beraud, Bulletin de la Soc. de Chir., 1862, Tom. III. p. 185.)

In the above case, reported by M. Richet, the trochanter was rotated towards the median line, with the head of the femur facing directly forward, and probably with displacement of the detached bone. But the fact that the round ligament was unbroken would seem to indicate that the luxation was only partial, as might indeed have been inferred from the position of the limb, which, though everted, was not much displaced.

In this connection, M. Richet (Ibid., p. 226) refers to a case of luxation of Maisonneuve (Rev. Méd. Chir., Tom. XVI. p. 48) in which a fragment of a broken acetabulum had in twenty-seven days united with the rest of the rim so firmly that the fracture could hardly be discovered.

tion of the force which has inflicted the injury, and perhaps in obedience to circumstances not always to be explained."[1]

In two of the recorded cases the patients recovered, being able to walk; in one of these the head of the femur had become almost completely inclosed in a bony shell. In two other cases, the patients died of the injury, which in all was the result of great local violence.[2]

It may be remarked, that, when the head of the femur is thus thrust completely within the pelvis, the capsule and surrounding muscles are relaxed, and would not determine the position of the bone.

ASSERTED FRACTURE OF THE ACETABULUM, WITHOUT CREPITUS, FROM A SUPPOSED IMPOSSIBILITY OF KEEPING THE FEMUR IN PLACE.

It has been already remarked that the evidence in this class of cases is unsatisfactory, and it is not unlikely that the bone could have been kept in place by angular extension, when other means had failed, or by confining the leg in the position of the final manœuvre by which it was reduced, as before described.[3]

[1] Op. cit., p. 343.
[2] In the case of Lendrick, and that of Morel-Lavallée, the accident was supposed to be that of fracture of the neck, from which it may be inferred that the foot was everted. In Case LXXII. of Cooper, the appearance was that of dislocation backward, probably involving inversion. In that of Moore, the limb was shortened two inches, slightly flexed and abducted, but without rotation in either direction. Cooper, Op. cit., Cases LXXII. and LXXIII.; Lendrick, Amer. Jour. Med. Science, August, 1839, Vol. XXIV. p. 481 (from London Med. Gaz., March, 1839); Morel-Lavallée, Malgaigne, Op. cit., Tom. II. p. 881; Moore, Med. Chir. Trans., 1851, Vol. XXXIV. p. 107.
[3] See p. 56. In the case of Keate (Cooper, Op. cit., Case LXIX.),

Fracture of other Parts of the Pelvis.

A fracture of the pelvis, not especially involving the acetabulum, can hardly be mistaken for luxation of the hip; and yet the following case, under my care, may be cited as an instance of a limb the position of which, when first seen, was identical with that of a dislocation, and, as in similar cases, was probably due to an effort of the patient to relieve the pain of injured tissues.[1]

The patient, a young man of seventeen years of

---

the fact that the limb could be drawn down, together with doubtful crepitus, was regarded as evidence of fracture of the socket.

For a case of Mr. Brodie, of twelve weeks' standing, where failure to reduce a dorsal dislocation was attributed to fracture of the socket, although none of its indications were present, see Lancet, Vol. XXIV. p. 671.

The following case of supposed fractured socket without crepitus is one of several reported by M. Richet. A young man fell, in dancing, while endeavoring to fling up his leg to the level of his partner's face. The leg was much inverted, and three quarters of an inch shortened, the head of the bone being felt upon the dorsum. The bone was repeatedly reduced, and as often escaped. The patient was ultimately placed in a fracture apparatus with extension, and two years after walked lame, the head of the bone rising upon the ilium at each step. No crepitus was felt, the diagnosis being based upon the supposed impossibility of keeping the head in the socket. (Bull. de la Soc. de Chir., 1862, Tom. III. p. 251.)

[1] A case of fracture of the ilium yielded crepitus under pressure upon the anterior and upper part of the ilium, the leg being shortened three quarters of an inch, and the foot slightly everted. After extension by the double inclined plane for several weeks, the deformity disappeared. (Lancet, Vol. XLIV. p. 877.)

In a case of fracture of the ilium, the right leg was half an inch shorter than the left, and slightly everted, with flattening of the region of the trochanter, the knee being also abducted. Pressure on the anterior superior spine produced crepitus attended with acute pain in the joint. (Lancet, Vol. XV. p. 575.)

age, entered the Massachusetts General Hospital, having been caught beneath a heavy piece of machinery, which fell from a wagon, striking upon the front of his left thigh just below the groin. Upon examination, the thigh was found to be flexed upon the pelvis, and the foot everted. The knee was widely separated from the other, any attempt to approximate them causing pain. The pubes was tender, when pressed. Under ether the leg resumed its normal position. No crepitus was discovered, although the patient had complained of a sense of grating in the perinæum. A broad strap was placed around the pelvis, and in six weeks the patient was well enough to be discharged, walking on crutches. It is difficult in this case to account for the position of the limb before etherization, except on the supposition that it may have afforded relief to pain. To the eye, its position was that of a thyroid luxation.[1]

[1] Mass. General Hospital Records, Vol. CXXVII. p. 210.

www.ingramcontent.com/pod-product-compliance
Lightning Source LLC
Chambersburg PA
CBHW030357170426
43202CB00010B/1404